RAILWAY · TIMES ·

Contents

Introduction	3
Blue is no longer the colour	5
A Vulcan in a culvert	6
Tollesbury Termination	9
The Mikado that never did exist	12
Sandling - a junction no more	16
Two New Festival of Britain Ventures	19
Taking stock at Eardley	21
Sir Eustace calls it a day	26
Signs of yesterday's times	28
Rule Britannias! First year operations and experiences	31
Report on Railway Electrification	34
This is York - 1951 style	39
Anglian 'Spam Cans'	43
North Sunderland succumbs	45
Newish (!) 4-car EMU for the Southern Region	51
Last H1 Atlantics withdrawn	53
Let's Halt Awhile	57
North of the Border in '51	60
Golden Arrow Pullmans & the 'Trianon' Bar(s)	63
From a Waterloo Eyrie	67
Crane tank bows out at Bow	68
Rail Transport at the Festival of Britain	69
Brief Encounters	76
In the next issue covering British Railways in 1952	80

RAILWAY TIMES 1951

Front cover: The introduction of the new Britannia pacifics on the prestigious 'Golden Arrow' Pullman express is epitomised in this view of No. 70004 *William Shakespeare* at the head of the down service in The Warren near Folkestone. Allocated to Stewarts Lane from October 1951 until the summer of 1958, No 70004 was a regular and a routinely immaculately presented performer on the 'Arrow'.

Above: One of the many branch line casualties of 1951 was the route between Kelvedon and Tollesbury in Essex whose rustic impecunious charm is captured in this view of Tolleshunt Knights station. What one writer described as "a particularly fine early GER coach specimen", a 4-wheeled 5-compartment third class vehicle dating from 1872, constitutes the waiting accommodation here, although by April 1951, the date of this view, it had undoubtedly seen better days. *NS207371A / Transport Treasury*

Rear Cover: The interior of the new Trianon Bar Pullman car *Pegasus*. Bar stools were not provided and drinkers had to perch rather uncomfortably against the rail if they did not wish to partake of the narrow bench seat on the right. The striking marquetry panel seen in the background and taken from car No 5 depicted English oak leaves and French vine leaves representing the 'Entente Cordiale'.

Copies of many of the images within **RAILWAY TIMES** are available for purchase / download.
Wherever possible illustrative material has been chosen dating from 1951 or is used to reflect the content described.

In addition the Transport Treasury Archive contains tens of thousands of other UK, Irish and some European railway photographs.

© Jeffery Grayer. Images (unless credited otherwise) and design The Transport Treasury 2024

ISBN 978-1-913251-80-2

First Published in 2024 by Transport Treasury Publishing Ltd.,
16 Highworth Close, High Wycombe, HP13 7PJ

www.ttpublishing.co.uk or for editorial issues and contributions email to **admin@ttpublishing.com**

Printed in the Malta by the Gutenberg Press.

The copyright holders hereby give notice that all rights to this work are reserved.
Aside from brief passages for the purpose of review, no part of this work may be reproduced, copied by electronic or other means, or otherwise stored in any information storage and retrieval system without written permission from the Publisher.

This includes the illustrations herein which shall remain the copyright of the respective copyright holder.

Introduction

With this the fourth volume of Railway Times we have reached the year 1951 when the prestigious Festival of Britain was opened. Some of the transport elements displayed on the South Bank site in London are featured in this issue.

In addition to covering as usual some of the more important developments of the year on BR we also take a look at some of the lesser regarded sights, common at the time, such as signage often of pre-Grouping companies, and a selection of rustic halts which have now passed into history.

Coaching stock, an oft neglected element of the railway scene, is highlighted in a couple of articles featuring brand new Pullman rolling stock and those vehicles coming to the end of their working lives.

Continuity, in the shape of Sir Eustace Missenden, the last General Manager of the Southern Railway and latterly the Chief Executive of BR, came to an end with the announcement of his retirement from the railway scene.

The Railway Executive decided that the blue livery adorning express locomotives was to be dropped in favour of the more traditional green.

The new Britannia locomotives made their presence felt on several regions although early problems were experienced resulting in the transfer of several Bulleid pacifics to the Eastern region whilst the problems were rectified.

Branchline closures continued and we spotlight three more casualties from Sussex, Essex and Northumberland. The number of locomotive classes was also being inexorably thinned as evidenced by the withdrawal of the last of the famous Brighton H1 Atlantics and the retirement of BR's oldest serving locomotive at Nationalisation, the ex LMS crane tank from Bow in London.

Our regular feature 'Brief Encounters' as before highlights some of the quirkier happenings on the network.

I hope you find something of interest within the varied pages of Railway Times No 4.

Editor: Jeffery Grayer

No. 60027 *Merlin* wears its BR blue livery in this view taken at Edinburgh's Haymarket shed in 1951. In July 1950 No 60027 had received the new BR dark blue livery with white lining and the 'cycling lion' emblem on the tender that had been introduced earlier that year. As mentioned in one of the articles in this volume the BR blue did not wear well in service and the standard livery was changed to Brunswick green with orange and black lining. This was applied to *Merlin* in June 1952 and was retained until withdrawal in September 1965. The naval plaque seen on the side of the boiler cladding had been fitted in 1946 and this too remained in place until the end.

RAILWAY TIMES 1951

On what looks like a cold frosty day at Haymarket, No 60066 *Merry Hampton* is well-coaled and ready for its next assignment whilst decked out in the short lived BR dark blue livery which it received in October 1949

The 'Bournemouth Belle' prepares to leave Bournemouth West with blue liveried Merchant Navy No 35015 Rotterdam Lloyd at its head. *S C Townroe*

Blue is no longer the colour

With apologies to Chelsea FC..........

The colour blue has been deemed no longer suitable by the Railway Executive for BR's express passenger locomotives. After a period in use it was found to wear badly and when existing stocks of paint are exhausted it will be abandoned. Blue was adopted in 1949 after the general public were invited to review a variety of colour schemes. Whilst blue was readily acceptable to devotees of the LNER and LMS, blue having been used by both companies in the past, it was perhaps not quite so welcomed by GWR or SR enthusiasts for many of whom the sight of a blue King or Merchant Navy was anathema. Many will undoubtedly welcome the return of green livery. The following images depict that short period when blue really was the BR colour of choice.

In this 1950 view blue livered King class No 6020 *King Henry VI* halts at Leamington Spa. No 6020 was the first of the final batch of ten locomotives built at Swindon in 1930 and is carrying a Wolverhampton shedplate (86A) where it would reside for the remainder of its career until withdrawn in July 1962.

Fresh out of the Doncaster paintshop and newly named, *Bon Accord,* A1 pacific No 60154 sports blue livery and is waiting to be turned on the triangle at Doncaster shed in April 1951. Like many of its contemporaries this example had a relatively short career of just over 16 years.

A Vulcan in a culvert

No not Mr. Spock in the gutter but rather an 0-6-0 taking the plunge!

Class C2X and their predecessors the C2 class were known colloquially as 'Vulcans' having been constructed at the Vulcan Foundry in Newton-le-Willows.

No 32522 which had been hauling the 9:30am goods service from Chichester to Midhurst plunged down the embankment which had been washed away by water impounded following the blocking of a culvert some half mile south of Midhurst station. Fortunately driver Fred Bunker of Hove was alerted by his passed fireman George Howes of Bognor who had noticed the yawning chasm and both managed to jump clear.

Breakdown trains from both Brighton and Fratton were brought to the scene but were unable to retrieve the locomotive or tender due to the instability of the embankment adjacent to the collapsed culvert but they were able to remove the wagons.

It was only after several hundred yards of embankment had been removed towards the station to the south at Cocking that gangs were able to lay some temporary track to enable the tender and locomotive to be hauled up the incline using Kelbus gear. The near vertical position of the locomotive tender still attached to one wagon of its goods train is apparent in the above view, the remaining wagons having been withdrawn by this time. It was the weight of the tender and of the locomotive on an unstable embankment that precluded an easy recovery.

The resulting gulley caused by the collapsing culvert was some 30 feet deep. The 3 ton contents of the tender promptly caught fire as it cascaded down to the open firebox door and the conflagration was replenished by a further 10 tons of coal being gravity fed from the smashed wagon. This was left to burn itself out over several days and was still smouldering some three weeks later. It was not until the end of February that recovery went ahead and the C2X taken to Brighton Works for repair. Despite the fact that it was then 51 years old the 0-6-0 was repaired and coupled to a replacement tender in the shape of one cobbled together from that of another C2X and a scrapped T9 class. No 32522 saw a further ten years of service until withdrawn from traffic in October 1961.

RAILWAY TIMES 1951

Images right: Passenger services on the line between Midhurst, Cocking, Singleton, Lavant and Chichester had ceased back in 1935 after a life of just 51 years. Local freight continued to use the line for some years - munition trains for Portsmouth being stabled until needed in the line's tunnels in WW2.

Due to its infrequent and light traffic loads the collapsed culvert was never repaired after the accident and goods on the remaining section was slowly cut back to Lavant surviving until the 1970s.

Bottom; Work stained but seemingly none the worse for its plunge, sexagenarian No 32522 was photographed on Three Bridges shed on 3 June 1960.

Further images of this incident are contained in the Editor's book 'Rails along the Rother' published by Transport Treasury in 2024..

RAILWAY TIMES 1951

Tollesbury Termination

The Kelvedon and Tollesbury Light Railway was a locally promoted railway company intended to open up an agricultural district that suffered from poor transport links.

The line opened from Kelvedon to Tollesbury in 1904. At Kelvedon the railway had its own station close to the GER main line station. All its stations had minimal buildings with in most cases an old coach or bus body serving as waiting rooms whilst the passenger rolling stock consisted of conversions of old vehicles. Passenger business was never brisk but the area around Tiptree experienced major growth in the culture of soft fruit and production of jams.

In its latter years passenger journeys on the line had reduced to between eight and ten persons daily - clearly financially a hopeless case as far as passenger traffic was concerned.

Accordingly, the last passenger journeys ran on 5 May 1951 headed by class J69/1 No 68578 with smokebox wreath attached. This was followed by the withdrawal of goods facilities beyond Tudwick Road Siding, to the south of Tiptree, from 29 October 1951. The traffic on the residual branch consisted of coal inwards and Tiptree jam and preserve products outwards. This arrangement continued for some years until the final revenue earning run on the branch was made on 28 September 1962.

Opposite top: Close up of the ancient rotting coach body which was an ex Eastern Counties Railway 4-wheeled saloon withdrawn in the 1880s. This constituted a lock up store at Tolleshunt D'Arcy station. The name Tolleshunt derives from the Anglo Saxon meaning 'Toll's Spring' whilst D'Arcy reflects the original gift of the estate from William the Conqueror to the family of the same name. *Lens of Sutton, Dennis Culum collection 0956*

Opposite bottom: An April 1951 image shows the layout at Tollesbury station looking towards the pier. The tall water tower seen in the background is not railway related being part of the local waterworks infrastructure. The entry points to the goods yard are in the right foreground and the coach body on the platform acts as parcels store and lamp room. *Neville Stead 207373 / Transport Treasury*

Above: Typical of the passenger rolling stock used on the line was this vehicle an old GER six wheeled brake third coach No. E62262 seen here at Tollesbury station. *Lens of Sutton, Dennis Culum collection 0957*

RAILWAY TIMES 1951

Left: The writing is literally on the wall for the branch line. This closure notice posted on an LNER headed board on Tollesbury station indicates the withdrawal of passenger services on and from Monday 7 May 1951. *Neville Stead 207373B / Transport Treasury*

Bottom: Tolleshunt Knights witnesses the arrival of Colchester based Class J69/1 No 68578 on 23 April 1951 with a mixed train. The origin of the village's unusual name may lie either with the fact that the manor once belonged to the Le Chevalliers, being the Norman French for knight, or that the surrounding land was once controlled by the Knights Templar who had considerable holdings in Essex until Edward II suppressed the order in 1311. *Neville Stead 207371 / Transport Treasury*

RAILWAY TIMES 1951

Another of the intermediate halts was Inworth seen here from the train where an old coach body did duty as a passenger shelter. *Lens of Sutton, Dennis Culum collection 0955*

As with many a country branchline, its nemesis was in the shape of a more convenient bus service. In the case of the Tollesbury Railway this was provided by the firm of Osborne one of whose Bedford OBs, EX 6666, is seen here in Colchester in September 1960 on a service from Tollesbury. What price such a numberplate today? *JSCPA030*

Operating for more than a century, the firm of Osborne initially used horse drawn vehicles for a Tollesbury based passenger and parcel service. They then started a passenger service between Tollesbury and Colchester in 1917 using a Model T Ford and in the 1920s a school bus service was introduced to Maldon and later a regular bus service was introduced to the town. In 1997 Osborne sold out to the Hedingham Bus Co. today known as Hedingham & Chambers part of the GoAhead group and still serving the area. This view of a Daimler CVD6 with Willowbrook B35F body and registered as NN 0310 was taken as it climbed Market Hill in Tollesbury en route to Maldon. *On-Line Transport Archive 0055*

The Mikado that never did exist

*"A more unlikely Mikado never
Did in Britain exist"*

(With apologies to Gilbert & Sullivan)

One of the 12 new Standard designs produced by BR in their 'Report on proposed Standard steam locomotives' published in 1948 was intended to be a 2-8-2, this wheel arrangement being known as the 'Mikado' type. The name had originated from a group of Japanese type 9700 2-8-2 locomotives that were supplied by the Baldwin Locomotive Works for the 3' 6" gauge Nippon Railway of Japan in 1897.

In the 19th century the Emperor of Japan was often referred to as 'The Mikado' in English and the eponymous Gilbert & Sullivan opera which premiered in 1885 achieved great popularity in both Britain and America.

In 1951 details of this new Mikado were sketchy with no details of cylinder size, axle load or tractive effort being supplied to the railway press which only had 'No particulars available yet' to report.

The 2-8-2 wheel arrangement although rarely employed on UK railways had been notably adopted by Gresley in two of his LNER designs namely the P1 and P2 classes.

The P1 was a freight derivative of his more famous A1 class but only two were constructed as there was little need for their abilities and they remained rather under-utilised throughout their short existence.

His other class, the six P2s, were express passenger locomotives built to haul heavy express trains in the hilly terrain north of the Scottish capital to Aberdeen where Gresley considered that the additional adhesion possible with this wheel arrangement might prove useful.

Unfortunately they turned out to be the wrong horse for this demanding course. The long rigid wheelbase of the P2s was unsuitable for the sharp curves north of Edinburgh and the locomotives were plagued with problems such as overheated bearings and the occasional broken crank axle. The track also suffered. Edward Thompson, Gresley's successor, later converted them to pacifics. In 2014 construction began on a new Class P2 Mikado locomotive, the *Prince of Wales* by the P2 Steam Locomotive Company.

LNER Class P1 No. 2393 seen here with Westinghouse brake pump fitted. The reduction in heavy freight trains after WW2 led to both engines of the class being withdrawn and scrapped under Thompson in July 1945. *Neville Stead 207060 / Transport Treasury*

Top: No 2394 has charge of an up coal train at Hadley Wood (2394)

Bottom: This undated view shows Class P2 No. 2001 *Cock O' The North*. Built in 1934 as a Mikado it was rebuilt as a pacific ten years later and under BR numbered 60501. However these rebuilds were never deemed a success and the class of six was withdrawn in the late 1950s and early 1960s.
Neville Stead 206548 / Transport Treasury

No. 503 *Lord President* wearing green livery was photographed at Edinburgh Haymarket depot in 1947. It had been rebuilt as Class A2/2 at the end of 1944 and would be withdrawn as No. 60503 at the end of 1959.

In 1906, the CME of the GWR Churchward had planned a class of Mikado tank locomotives to handle heavy coal trains in South Wales but the scheme was abandoned due to concerns that they would be unable to handle the sharp curves present on the Welsh branches.

Instead he turned his thoughts to designing the 42xx class of 2-8-0Ts of which nearly 200 were built. With the Depression of the 1930s coal traffic declined resulting in many of these locomotives standing idle as their limited operating range prevented them from being allocated to other mainline duties. Collett rebuilt 54 of them as 2-8-2Ts, known as the 72xx class, the addition of a trailing axle furthering the engine's operating range by allowing increased water and coal capacity.

The BR Standard design team led by Robert Riddles originally considered proposals for a heavy freight locomotive and Riddles, who had been responsible for the Austerity classes of 8F 2-8-0 and 2-10-0 designed and built during WW2, favoured a 2-10-0 wheel arrangement whilst his second in command E.S. Cox (Executive Officer Design), who did most of the actual design work, preferred a 2-8-2 arrangement.

Whilst the advantage of the 2-10-0 configuration allowed more of the weight of the locomotive to be available for adhesion its major disadvantage was the long coupled wheelbase, which despite the use of flangeless middle driving wheels, becomes problematic in goods yards and depots where the standard of track was often less than optimal leading to the increased likelihood of derailments.

Whilst a Mikado arrangement was initially contemplated by the design team, following Riddles own experience during his War Department service with German 2-10-0s which he had found impressive both in terms of simplicity and power output, he decided to design a BR equivalent instead of the anticipated Mikado. Apart from the pacifics, the Britannias, Clans and the sole example the *Duke of Gloucester*, the emphasis in the Standard repertoire was very much upon mixed traffic locomotives which led to work on a heavy freight locomotive being initially deferred.

In the book which Cox later wrote he detailed the proposal for a 2-8-2 using a 'Britannia' boiler, 5ft 3in diameter wheels, which gave an axle loading of just under 17 tons and a potential tractive effort of 35,912 lbs. He produced a memorandum outlining ten major advantages for this wheel arrangement. However Riddles successfully countered these arguments considering that in comparison with the existing 2-8-0 freight locomotives that it was designed to replace, it offered little in the way of increased adhesion. The wide firebox envisaged for the Mikado with a trailing carrying axle was also felt to mean that on starting off the locomotive would have a tendency to "sit back" with vital weight being thereby transferred from the coupled wheels resulting in a loss of adhesion at a critical moment, although this theoretical problem had never manifested itself in the operation of the earlier P1s and P2s.

As is well known Riddles and his team went on to design the very successful 9F 2-10-0 which did not appear until 1954 and thus the Standard Mikado was stillborn.

Before we leave the subject of Mikados that never existed, we should perhaps consider a design of that engineering maverick Bulleid who soon after joining the Southern Railway in 1937 turned his thoughts to new motive power to haul the heavy continental boat trains operating on the South Eastern section.

In June 1938 having originally considered a 4-6-2 design he then alighted on a possible 2-8-2 which he felt would give greater adhesion on the hilly Dover road. The Civil Engineer raised objections to the proposed leading pony truck and to overcome these Bulleid intended to employ the German Krauss-Helmholtz type in which the leading and carrying coupled axles are manufactured in the form of a bogie with suitable side play being allowed in the first section of the coupling rods thereby safely leading the locomotive into curves and over uneven trackwork.

Finding this acceptable the Civil Engineer cautiously authorised the construction of two prototypes. Bulleid was unhappy at this miserly commitment of only two feeling that he would be involved in endless arguments to get any further examples built. He thus abandoned this idea and after briefly considering a 4-8-4 came up with a pacific in the shape of the 'Merchant Navy' class and the rest as they say is history.

Class 72xx No 7201 is seen at Swindon displaying the 2-8-2 Mikado wheel arrangement but in tank locomotive form. *Peter Gray 3064*

Sandling - a junction no more

With the closure of the Hythe branch, Sandling station on the Margate to London line loses its junction status.

Top: Taken from high up on the cutting side showing the layout at Sandling with the Hythe branch curving away to the right and a pair of Camping Coaches in the yard.
Lens of Sutton 21766

Bottom: This view of the one remaining Hythe branch platform at Sandling Junction dates from 1950.
Lens of Sutton, Dennis Culum collection 0852

RAILWAY TIMES 1951

Having previously deposited its train at the platform No. 31521 engages in some light shunting in the grass grown goods yard. The yard contained a 30 cwt crane seen on the far right where a coal yard was also located. The sidings on the left had already been lifted by this date and in spite of having closed back in 1931 the platform mounted signalbox still survived twenty years later. When closure proposals for the passenger service were announced it was the original intention to continue to handle goods traffic at Hythe, the principal inbound freight being coal and stores for a local brewery with outbound goods consisting mainly of parcels traffic and empties from the brewery. However, with anticipated costs of £8,500 required in the near future according to the Civil Engineer's Department it was decided that this expense could not be justified and the line closed completely.
R C Riley 4284 / Transport Treasury

The decrepit state of Hythe station is testament to the fact that the branchline was on its last legs. The remains of the bomb damaged former Parcels Office seen to the right of the locomotive does nothing to enhance the scene. Having run round its two coach train, consisting of pull-push set No 721, Class H tank No 31521 is ready to return to the mainline at Sandling Junction on 25th August 1951. Hythe station was inconveniently sited high above the town and became the terminus of the branch following closure of the section onwards to Sandgate in 1931. This truncation rendered the branch without watering facilities so a water tower and crane were provided at the north end of the up platform at Hythe. *R C Riley 4290 / Transport Treasury*

Top left: Although Sandling had lost its junction status it still enjoyed patronage not just from travellers on the mainline but from holidaymakers who chose to stay in the Camping and Holiday Coaches that were parked at Sandling for many years as these views testify. Predecessor of the later pair of Pullman coaches was this vehicle an ex SE&CR third, formerly numbered 19325, seen here at Sandling on 8 May 1948. *Lens of Sutton 61752*

Top right: This pair of former Pullman coaches were stabled in the down siding in 1960 for use as camping coaches, access being retained along the former Hythe branch to serve the goods yard which remained open until February 1963. The nearer of the two vehicles was fitted with a full kitchen, two sleeping compartments and a room with two single beds. The vehicles, known as Pullman Holiday Coaches, were Car No 11 and Car No 16 both converted 8-wheel Pullmans (Now Nos P40/41).The platform road was retained as a refuge siding, and as a headshunt for the down siding where the two camping coaches were stabled, with the track remaining in situ for a further 15 years. *Lens of Sutton 61144*

Centre: Haine tunnel was the major engineering feature on the branch and this view from the northern portal shows the line, which was reduced to single track in 1931 after the extension to Sandgate was closed, continuing on to Hythe. *Lens of Sutton, Dennis Culum collection 0853*

Arriving at Hythe is Wainwright Class H 0-4-4T No 31521 on 21 August 1951. No passengers are in evidence - a not unusual state of affairs by this date. Housing now occupies the former railway site and trains to Hythe bringing holidaymakers to this part of the Kent coast are but a distant memory. However, tourists and locals alike can still experience train travel from Hythe albeit on the narrow gauge Romney, Hythe and Dymchurch Railway (RH&DR) which has its eastern terminus in the town. *R C Riley 4282 / Transport Treasury*

Two new Festival of Britain Ventures

'THE MERCHANT VENTURER' AND 'THE WILLIAM SHAKESPEARE' WERE INTRODUCED BY THE WESTERN REGION FROM MAY 1951

Carrying the appropriate headboard Castle Class No 5087 *Tintern Abbey* heads under Westbourne Bridge on the exit from Paddington with the down 'Merchant Venturer' heading for the West. *R C Riley 15232 / Transport Treasury*

'The Merchant Venturer' named train had its origin in the 1951 Festival of Britain when it was one of two services from Paddington designed to appeal to Londoners wishing to have a day trip to regional cultural centres hosting events as part of the Festival. In the case of 'The Merchant Venturer' serving Bath, Bristol and Weston-super-Mare with the other new service the 'William Shakespeare' serving Stratford-upon-Avon.

Although the latter ran for just one season, the 'Merchant Venturer' continued on a seasonal basis until 1961 a year after the June 1960 Restaurant Car tariff; illustrated right and overleaf. The name originated from the Bristol based Society of Merchant Venturers whose origins lay in the medieval guilds receiving their Royal Charter in 1552. It regulated and dominated sea trade through Bristol's important port for centuries becoming highly influential and wealthy.

Although today's Great Western Trains resurrected the name in December 2009 for an IC125 hauled London to Plymouth service via Bristol it, unsurprisingly in view of today's concerns with any organisation's past links to slavery, no longer appears in their timetables.

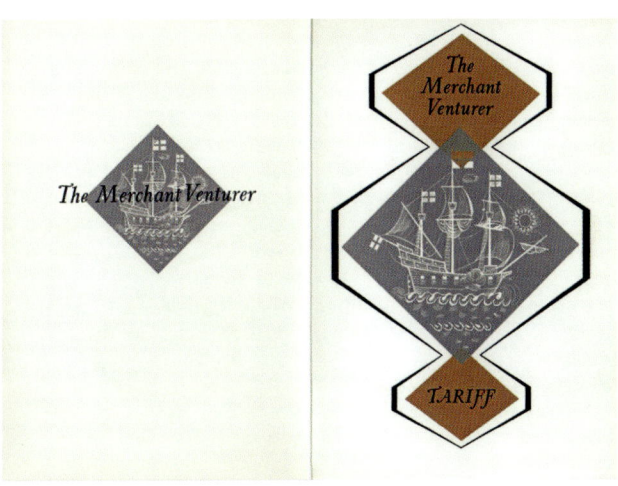

Morning Coffee is served when timings are suitable
Coffee per cup 10d.   Biscuits 6d.

A light refreshment and corridor service is made
available when practicable

LUNCHEON 11/-

AFTERNOON TEA 3/6

DINNER 12/-

Coffee 10d.

Children travelling at half-fare are charged reduced prices for
Table d'Hôte Luncheon and Dinner

It is particularly requested that a bill be obtained from the conductor for all payments. British Transport Catering Services desire to render every possible service to passengers and it will be appreciated if they will report any unusual service or attention to the Regional Catering Superintendent, Western Region, British Transport Catering Services, Paddington Station, W.2. Complaints, to which the Conductor's attention should be drawn, will be investigated and a remedy sought. In the general interest, passengers are kindly asked to refrain from smoking immediately before or during the service of meals.

WINES

When you are ordering luncheon or dinner on this train, may we draw your attention to the interesting wines offered at prices as reasonable as any you will find in this country.

SHERRY				LIQUEURS		
Medium Dry	Glass	2/6		Van der Hum	Miniature	4/9
Amontillado No. 4, Pale Dry		2/6		Cointreau		4/3
Fino No. 7, Pale Dry		2/6		Bénédictine		3/9
Walnut Brown		2/6		Drambuie		3/9
APÉRITIFS				Bolskümmel		3/6
Gin and Lime, Orange or Lemon		2/6		Cherry Heering		3/-
Gin and Bitters		2/3		SPIRITS		
Gin and Vermouth, French or Italian		2/9		Cognac Vieux Maison,		
Vermouth, French or Italian		1/9		30 years old	Measure	3/-
Tomato Juice Cocktail	Baby Bottle	1/-		Brandy***		3/-
Pineapple Juice Cocktail		1/-		Gin		2/3
BORDEAUX Red				Rum		2/3
	Bott.	½ Bott.	¼ Bott.	Whisky—G.W.R. Special		2/6
Médoc	13/-	7/-	3/9	Whisky—Proprietary Brands		2/6
Château Mouton				Whisky—Proprietary Brands	Miniature	5/-
d'Armailhacq 1953	20/-	10/6	—	Brandy		5/-
BORDEAUX White				Gin		4/6
Graves	12/-	6/6	3/6	Rum		4/6
Sauternes	15/-	8/-	—	CORDIALS AND FRUIT JUICES		
BURGUNDY Red				Lime Juice	Glass	7d.
Mâcon	13/-	7/-	3/9	Lemon Squash		7d.
Beaune	20/-	10/6	—	Orange Squash		7d.
BURGUNDY				Grape Fruit Squash		7d.
Vin Rosé	13/-	7/-	—	Apple Juice	Split	1/-
CHAMPAGNE				BEER, STOUT AND LAGER		
G. H. Mumm,				Bass, Worthington	Bottle	1/7
Cordon Rouge 1952	47/6	24/6	—	Double Diamond		1/7
Perrier-Jouët N.V.	40/-	20/6	11/-	Other Proprietary Brands		1/7
ALSATIAN				Guinness		1/7
Sylvaner	16/-	8/6	—	Mackeson's Stout		1/7
				Whitbread's Pale Ale		1/4
SPANISH				Other Light Ales		1/3
Spanish Graves	10/6	5/6	3/-	British Lager		1/9
Spanish Burgundy	10/6	5/6	3/-	Tuborg Lager		1/9
SOUTH AFRICAN				Carlsberg Lager		1/9
Paarl Amber Hock	10/6	5/6	3/-	CIDER		
AUSTRALIAN				Cider	Bottle	10d.
Emu Burgundy	10/6	5/6	3/-	Champagne Cider	Reputed Pint	3/6
PORT				MINERALS		
Tawny	Glass	2/6		Schweppes' Aerated Water	Baby 6d.	Splits 8d.
Very Fine Old		3/-		Apollinaris	6d.	9d.
				Vichy Célestins		1/9
				Ginger Beer	Bottle	9d.

CIGARS · CIGARETTES · CHOCOLATES

Table 2

THE MERCHANT VENTURER
RESTAURANT CAR SERVICE

LONDON, BATH SPA, BRISTOL
and WESTON-SUPER-MARE

WEEK DAYS

		am				pm
London (Paddington)	dep	11A15	Weston-super-Mare General	dep		4A35
			Yatton		"	4 47
			Nailsea and Backwell		"	4 56
		pm	Bristol (Temple Meads)		"	5A27
Bath Spa	arr	1 1	Bath Spa		"	5 48
			Chippenham		"	6 10
Bristol (Temple Meads)	"	1 22	Swindon		"	6 40
			Reading General		arr	7 21
Weston-super-Mare General	"	1C56	London (Paddington)		"	8 5

A—Seats can be reserved in advance on payment of a fee of 2s. 0d. per seat (see page 23).
C—On Saturdays arr 2 4 pm

The tariff choices reflects the timing of the service i.e. an 11:15 am departure from Paddington and a 4:35pm return from Weston, with luncheon, afternoon tea and dinner available in the Restaurant Car. The wide range of wines and spirits offered included the still available GWR Special Whisky blend at 2/6 (25p) some 12 years after nationalisation! Overstocking or unpopularity? *Tariff images courtesy of Mike Ashworth*

By the 1958 timetable, a page from which is illustrated left, the return service had several additional stops leading to a journey time of 3½ hours from Weston in comparison to the outward run taking just 2 hours 41 minutes.

As previously mentioned 'The William Shakespeare' ran for a single summer, from 3 May to 8 September 1951. Both the new trains utilised the new Mark 1 coaches painted in British Railways carmine and cream livery. 'The William Shakespeare' operated from Paddington to Wolverhampton, with a through portion of four carriages attached and detached at Leamington Spa for Stratford-upon-Avon. The main section left London at 10.10am and returned from Stratford and Wolverhampton at 7.23pm and 7.50pm respectively. Unusually for a named express the train was hauled on some occasions by Class 51xx tank locomotives although only on the Leamington Spa to Stratford-upon-Avon part of the journey. Although intended as a summer dated service it began earlier to coincide with the opening of the Festival at the start of May. Due partly to poor promotion, it did not appear in the main timetable which commenced two months later than usual in July that year due to the austerity measures still in force, the service was unsuccessful and only ran for four months in 1951.

Taking stock at Eardley
Captions courtesy of Mike King

Eardley Road sidings were situated beside the ex-LBSCR line from Streatham to Mitcham Junction and occupied a large area of level ground to the south east of the railway. Access to the South Western section was possible via Tooting and Wimbledon, the South Eastern section was reached via the Tulse Hill to Herne Hill connection at Peckham Rye and at London Bridge while all areas of the Central section could also be accessed. Stock could therefore be berthed here for any south London service regardless of originating station. However, it is probably fair to say that relatively few South Western line trains originated here and those that did were more likely to be occasional services such as stock for first class race trains and the like.

The following images cover some of the interesting coaching stock that could be seen here during 1951. The sidings remained in use until the effects of the Kent Coast Electrification were felt.

After the introduction of EMUs on the Kent services fewer and fewer vehicles were located here with many simply rotting away before formal withdrawal. During WW2 a number of the coaches stored here had been damaged by enemy bombing in both 1941 and 1944. Today the site is occupied partly by housing while a portion was designated as a local nature reserve by the London Borough of Lambeth in 2022.

This view, looking south from the vantage point of Streatham Junction South signalbox, reveals the extent of the carriage sidings at Eardley. *Lens of Sutton, Dennis Culum collection 1348*

An SE&CR 'Continental' corridor first, to SR Diagram 496. This is coach S7380 which was a 1924 vintage vehicle built by the Birmingham Railway Carriage & Wagon Co. It is marshalled in a rake of similar first-class coaches so probably forms a South Eastern or Newhaven boat train set. The livery is malachite green just with 'S' prefixes to the numbers but otherwise still in Southern Railway style. Compare the two styles of door grab rails. Those on both sides of the doors on coach 7380 are the original straight pattern, while those on coaches to the left and right are "kinked" and are fitted to the right hand side of the door openings only. This indicates a modification that began to be introduced in about 1950 following an incident where a passenger fell from a moving train. The original design used inward opening doors – possibly satisfactory for boat train services but when the coaches began to drift on to special and excursion work this became a potential problem. Most coaches were modified before withdrawal and the SR diagram number then had an A suffix to indicate this. Coach 7380 was downgraded to an all third in 1954 being renumbered as S647S and lasted in service until 1958. *Lens of Sutton, Dennis Culum collection 1028*

RAILWAY TIMES 1951

In use as a shunter's cabin is this former LB&SCR 26ft Stroudley four-wheeled, five-compartment third. It was built in 1885 as LB&SCR No 229 and was withdrawn in November 1921 and most likely grounded here soon after. It would have probably spent its whole life on suburban services out of Victoria and London Bridge as part of a set of anything up to 14 vehicles. As one of over 600 similar vehicles built between 1872 and 1891 it was easily the most numerous type of Brighton passenger coach. Just 41 examples survived to become Southern Railway stock as third class Diagram 57 at the Grouping, with two more as second class to Diagram 241, but all except a handful had been taken out of service by the end of 1925. Two were sent to the Isle of Wight while a few others became part of the Lancing Works staff train and these survived into the 1930s. Many more were grounded, both on and off railway premises and a few of the latter have been acquired more recently by the Bluebell and other heritage railways for preservation. To date, only a couple have progressed towards restoration. Note the addition of a corrugated iron roof and a ship-lap timber extension at the left hand end – either a store or a 'privvy'. *Lens of Sutton, Dennis Culum collection 1025*

This is SR Diagram 164A ex-SE&CR 'Continental' brake third No S3549 – one of 12 completed by the Metropolitan Carriage, Wagon & Finance Company in 1924 for South Eastern section boat train services. Use on these services plus the vertically matchboarded lower bodysides ensured their nickname 'Continentals' for their entire life. Indeed, apart from the war years, it was only after Nationalisation that they could be found on other services. One is preserved (No 3554) on the Keighley & Worth Valley Railway while one other somewhat modified departmental survivor still exists on the Kent & East Sussex Railway. Dimensionally these coaches were 62ft long over body and 8ft 6¾in wide. Similar coaches had been built by the SE&CR at Ashford Works since 1921 but these were six inches narrower, so were to SR route restriction '0', meaning that they could traverse the Tonbridge-Hastings line. All those vehicles ordered after the Grouping were to route restriction '1'. They were also the first SE&CR coaches to be equipped with buckeye couplings and Pullman-type gangways – unseen at the right-hand end - but this did not include the guard's brake end seen here which had no gangway and possessed screw couplings instead. Withdrawal of this coach came in July 1959 having been made redundant by the completion of Phase 1 of the Kent Coast electrification. *Lens of Sutton, Dennis Culum collection 1026*

RAILWAY TIMES 1951

This is a Maunsell 'Thanet' restriction '1' coach dating from 1924/25 – the first new design produced by the Southern Railway after the Grouping. There was a severe shortage of corridor stock on lines to North Kent and the 'Continentals' were not really suitable – having few access doors – so a new design was called for. Maunsell and Lionel Lynes took the L&SWR 'Ironclad' stock and produced an equivalent 57ft long vehicle but six inches narrower to enable them to be used on the South Eastern section. Ashford standard details and underframes replaced the Surrey Warner features of the 'Ironclads'. Number S3569 is an example of the brake third – to SR Diagram 165 – with five compartments, a lavatory and brake/luggage compartments. In this view we see the corridor side of this vehicle which formed part of set 461 – at that time an 8 coach formation, still used regularly on services to Ramsgate and Margate. British standard gangways and screw couplings were provided, the last new SR designs to have these features. *Lens of Sutton, Dennis Culum collection 1029*

The final generation of Southern Railway stock is represented by Bulleid semi-open brake third No S2521 to Diagram 2123 and brand new at the time of this photograph. This was at one end of 3-coach set 860 with composite S5918 and brake third S2522 forming the other vehicles. Note the destination board 'Waterloo'. BR crimson lake and cream livery is now carried, complete with Gill Sans numerals below the waistline. The Southern Region management were never keen on this livery being difficult to keep clean and which did not lend itself to the Southern's habit of regular patch repainting and revarnishing and so they saw to it that BR Southern Region green livery was reinstated as soon as possible after mid-1956. These Bulleid coaches were 64ft 6in long over body – some 5ft 6in longer than the previous Maunsell (and a few early Bulleid) coaches. The set ran in this form from May 1951 until November 1965. From the left, the coach comprises a luggage van, guard's compartment, two third class compartments served by a side corridor, a lavatory, a cross-vestibule, a 48 seat four bay third class saloon with central aisle and another cross-vestibule at the right-hand end. *Lens of Sutton, Dennis Culum collection 1030*

Above; Something a bit different as here we have ex-L&SWR post office sorting van No S4906, originally to SR Diagram 1184 – both numbers being in the van stock section of the listings – in malachite green with the number prefixed 'S'. Note the late fee letter box on the right hand bulbous portion of the bodywork. For an additional fee of one half-penny, the public could post letters into this box right up to departure time of the mail train. The van's regular duty had been the Waterloo to Dorchester mail train, due to leave Waterloo around 10.50pm each evening (with a similar up departure time from Dorchester) but probably by 1951 newer vans of Maunsell vintage had taken over and this reduced it to being a relief vehicle. Despite this it appears in exemplary condition. It was built in 1900 and survived until 1955, although some changes were made to the vehicle in its later years – the diagram number then being amended to 1209. Dimensions were 44ft long over body and 8ft 6in wide. The van is parked between two other interesting vehicles – on the left is generator van 97s (which ran with Bulleid sleeping car 100s) and on the right van 1309s which was the generator van for cinema car 1308s. Note also the wooden-centred Maunsell wheelsets. *Lens of Sutton, Dennis Culum collection 1034*

RAILWAY TIMES 1951

Opposite bottom: And here is another interesting vehicle in the shape of inspection saloon DS1. This was a former L&SWR arc-roofed coach dating from as far back as 1885. It saw several modifications during its long career including a new, well second-hand, underframe from another L&SWR corridor coach in 1950. This enabled it to continue in use until withdrawal in September 1963. It could occasionally be glimpsed going about its inspection duties often being propelled by a locomotive to give the occupants a better view of the track and during 1959-61 it was much in evidence on the Kent Coast lines while crew training for the forthcoming full electrification was in progress. It was finally broken up in October 1964. The livery is lined crimson lake, as befitting its non-corridor coach status, with the original 'left-hand end' numbering. Dimensions were 46ft long and 8ft wide. *Lens of Sutton, Dennis Culum collection 1035*

Above: Alongside DS1 was the former LB&SCR inspection saloon, numbered DS291. Altogether more modern and palatial, this was 61ft long and was completed during World War 1 for the great and good of the Brighton line to carry out their inspections (as LB&SCR No. 60). It ran on two 6-wheeled bogies and was somewhat rebuilt in 1934, again enabling it to continue in occasional service until June 1965. Being a corridor coach it has received crimson lake and cream livery but ended its days in BR Southern Region green. It was then purchased by the Bluebell Railway and, in the early days of preservation, it was used especially to serve cream teas on Sunday afternoon trains! After this period of activity it was taken out of use and remains stored until the present time with required repairs estimated to cost in excess of £100K so an imminent return to service is considered unlikely. *Lens of Sutton, Dennis Culum collection 1036*

In this Festival of Britain year a bronze plaque manufactured at Swindon Works was unveiled near the down platform at Talyllyn Junction near Brecon on 23 May.

Sir Eustace calls it a day

Sir Eustace Missenden Chairman of the Railway Executive has decided to retire at the end of January 1951 after three years in post. Five different executives reported to the BTC one of which was the Railway Executive headed by Missenden

Eustace James Missenden born in London and the son of a SE&CR station master started work as a junior clerk with the SER in May 1899.

His railway career was spent largely with the Southern Railway where he rose to become Docks and Marine Manager and in 1941 General Manager. Although a competent practical railwayman with a flair for organisation and delegation he apparently lacked the warmth and extrovert personality of Szlumper his predecessor at Waterloo, or the managerial distinction of Szlumper's predecessor Sir Herbert Walker whom he greatly admired.

He became the first Chairman of the Railway Executive of the newly formed British Transport Commission although he made it clear that he had only accepted the offer of the Chairmanship as a short term measure as it was his intention not to commit to a fixed period of tenure but to 'retire before too long'. This reluctance to remain in post was a reflection primarily of the fact that he did not move easily in Government circles, being suspicious of both politicians and civil servants. He found himself somewhat out of his depth in attempting to coordinate a team of disparate Railway Executive Members from the former 'Big Four' who were not responsible to him in the way that railway departmental officers in the past had been responsible to the General Manager of the SR. Difficulties stemmed from the fact that the Members each had their own agenda even when part of a nationalised industry and antipathy only increased when the Chairman came from the Southern the smallest of Britain's pre-nationalised railways.

Being a champion of electrification and dieselisation he found himself somewhat at odds with BR's policy of continuing to build large quantities of steam locomotives into the foreseeable future for in an address made in 1950 to the Institute of Transport he stated that "The steam locomotive inherently is inefficient and an extravagant user of coal. It is expensive to operate, service and maintain. Its operating characteristics, and the dirt and smell, prevent it from providing the quality of service required and expected by the public and which the railways must give if they are to survive".

He had been honoured in 1949 when Bulleid pacific No. 34090 was named *Sir Eustace Missenden - Southern Railway* reflecting both his own contribution to the war effort and that of the SR and its employees. This accolade was very much the brainchild of Bulleid and John Elliot, the latter had taken over from Missenden as acting GM of the SR. He was succeeded as Chairman of the Railway Executive by his former SR protégée Sir John Elliot. In 1953 the Executive was abolished and Sir Brian Robertson became Chairman of the British Transport Commission. Missenden died in Horsham aged 86 in 1973.

Missenden was not the only key figure to feature in the railway press this year which also marked the retirement of the last of the CMEs – George Ivatt and the death of Arthur Peppercorn whose recent retirement was featured in Railway Times No 1.

For students of armorial bearings the following is a description of the SR coat of arms – **Escutcheon:** Barry wavy of eight Argent and Azure four escutcheons in cross Argent that in chief charged with a sword erect Gules that in dexter a rose Gules barbed and seeded Proper that in sinister a torteaux thereon a leopard's face Or that in base a seahorse naiant (ie swimming) Sable. **Crest**: In front of a demi-sun Or a wheel Azure winged Argent struck through with a lightning bolt bendwise sinister Gules. **Supporters**: Dexter a dragon Gules sinister a horse Argent each resting its interior hind paw upon a wheel Azure. *Now you know!*

RAILWAY TIMES 1951

Above: No. 34090 was unique in many ways its nameplate being in three parts; in the length of its name and in having its plaque set on a horizontal rather than a vertical axis. The plaque which portrayed the Southern Railway coat of arms, nameplate and class plates were originally closely grouped towards the bottom of the air smoothed casing but were later more widely spaced over the full height of the casing. Whilst this unusual horizontal aspect suited the slab sided boiler casing of the original air smoothed pacific its original position low down on the casing certainly did not, as seen on display at the Ashford Works Open Day, 31 August 1949. *Arthur Taylor*

Right: Matters were perhaps slightly improved once No 34090 had been rebuilt. *BR2989*

Signs of yesterday's times

Cast iron signage, often featuring pre-Grouping company names, were a common feature of the railway scene in the 1950s. Here are just a few examples containing, often somewhat oddly, very specific information that one would not have thought justified the provision of such a sign.

Top left: M&GNR trespass sign located at the terminus at Yarmouth Beach *JH507*

Bottom left: At North Woolwich in 1951 this fascinating GER cast iron sign was found by the photographer mounted on an old coach body proclaiming a warning to 'Men working in sidings' regarding the use of red warning flags and lamps. *Lens of Sutton, Dennis Culum collection 0938*

Top right: In 1947 on the up side at Honor Oak station a LC&DR sign warns of the fact that the right of way will be closed very specifically on Good Friday and 'on such other days as the company may from time to time determine'. *Lens of Sutton, Dennis Culum collection 0112*

Bottom right: Staines in 1950 had two signs on one post – an L&SWR trespassing notice and a SR conductor rail warning. *Lens of Sutton, Dennis Culum collection 0564*

RAILWAY TIMES 1951

Top left: Pattenden siding located on the Hawkhurst branch between Goudhurst and Cranbrook still had, in 1951, a SE&CR sign declaring that 'No traction engine or other locomotive may be driven over this crossing until the permission of the nearest station master has been obtained.'
Lens of Sutton, Dennis Culum collection 1129

Bottom left: Another post that served a dual function was this specimen, found at Beddington Lane on the Wimbledon to West Croydon line, containing a Southern Railway warning against smoking and an LB&SCR trespass notice. *Lens of Sutton, Dennis Culum collection 2301*

Top right: This Midland Railway notice dating from 1899 was located at Grassington and threatened a penalty of £10 for trespassing upon the railway. In 1930, after only 28 years, the LMS had withdrawn regular passenger services due to poor patronage although excursion traffic continued for more than 30 years as did general goods traffic and stone traffic from the nearby quarry the latter often being hauled by Standard Class 4 4-6-0s. The end of quarrying in the early 1960s eventually led to the complete closure of the station and to the northern end of the branch in 1969. *Lens of Sutton, Dennis Culum collection 1666*

Bottom right: The K&ESR proudly displayed their notices in cast iron as did the larger railway companies and this example can still be seen at Tenterden Town. *JH2585*

RAILWAY TIMES 1951

Top: On 11 October 1951 No 70004 *William Shakespeare* is seen at Victoria prior to working the down Golden Arrow. *(Photo ID: 70004_Victoria_111051 © David P Williams Colour Archive)*

Left: No 70003 *John Bunyan* heading the down 'Norfolkman' leaves Liverpool Street on 29 September 1951. At that date the locomotive was just seven months old. *R E Vincent 58A 5 1 / Transport Treasury*

Opposite: No 70019 *Lightning* seen at Newton Abbot on 28 June 1952. *PG 0205*

Rule Britannias!
First year operations and experiences

The inaugural BR Standard locomotive No 70000, as yet unnamed, departed from its birthplace at Crewe Works with a dynamometer car train on 12 January in unlined black livery.

Undertaking running-in and preliminary tests in the Crewe area it also ventured on to the LMR mainline between Crewe and Carlisle. On 30 January and recognising a link with the Festival of Britain it was named *Britannia* by the Minister of Transport Alfred Barnes at a ceremony at Marylebone.

On 1 February it entered service with the Eastern Region, who will be receiving 15 of these Class 7 4-6-2s, hauling mainline trains between Liverpool Street and Norwich. The rest of the initial batch of 25 locomotives was earmarked for the Western Region although this did not turn out quite as planned with two, Nos 70004 and 70014 ending up initially with the Southern Region being joined by a third No 70009 for a brief period. Two, in the shape of Nos 70015 and 70016, also found themselves on the Midland Region. The names bestowed on these pacifics included some of relevance to the regions upon which they would be operating whilst some featured more traditional names.

EASTERN REGION. No 70000 operated the 10am down 'Norfolkman' to Norwich on 1 February and although achieving some impressive running developed a cylinder defect at Marks Tey on the return working on the following day necessitating some days spent in the repair shops at Stratford – not a very auspicious start.

The second Britannia was named by Lord Hurcomb Chairman of the BTC at Liverpool Street on 6 March. He said the new pacifics would be used on the continental services between Liverpool Street and Harwich and that when more came on stream reductions of between 10 and 40 minutes should be possible between Liverpool Street and Norwich.

No 70000 covered new ground in April when it worked the 11.05am from Cambridge to London. No 70000 was on exhibition in the north of England during May/June returning to London on 18 June with the service from Newcastle advertised as an advance portion of the 'Flying Scotsman' but which ran following the main service which was itself running ahead of time.

WESTERN REGION. Nos 70017, 70018 and 70019 were noted undergoing trials at Swindon in July and subsequently appeared on the 9am service from Swindon to Paddington and the 1.18pm from Paddington to Weston-super-Mare. The 10.45 Paddington to Cheltenham has seen No 70017 at its head with No 70018 noted on the 'Merchant Venturer'. The first Britannia to enter Cornwall was No 70019 when in August it worked the 'Cornish Riviera'. Nos 70020 and 70021 were both noted on service trains at the beginning of September.

No 70023 has been noted working between Paddington and Wolverhampton whilst No 70020 was observed working the 12.45am newspaper train from Paddington to Carmarthen

RAILWAY TIMES 1951

No 70009 at work with a regular service on the SR seen here entering Waterloo in 1951 with the 5:25pm departure from Southampton. *Lens of Sutton, Dennis Culum collection 0994*

No 70017 passes Vauxhall in 1953 during the Merchant Navy crank axle inspection period. *Lens of Sutton, Dennis Culum collection 1737*

as far as Cardiff being possibly the first occasion one has been seen in South Wales.

Following the failure of No 70004 on the SR, four of the WR pacifics were noted in store at Swindon in November pending a review of the cause, these being Nos 70018, 70021, 70023 and 70024. No 70024 returned to service at the end of December being observed operating between Swindon and Banbury.

SOUTHERN REGION

At the end of May No 70009 appeared on the SR hauling an Institution of Locomotive Engineers special to Eastleigh followed by several appearances on the 8.30am service from Waterloo to Bournemouth.

It was reported in the railway press that two Britannia pacifics Nos 70013 and 70014 were to be temporarily allocated to the SR at Nine Elms in exchange for Bulleid pacifics Nos 34039, 34057 and 34069 which were to be loaned to the ER. (Ed: See article in this issue 'Anglian Spam Cans'.) Later information reported that Nos 70009 and 70014 were at work on the Western Section although No 70013 will not now be joining them with No 70009 working the Bournemouth Belle on occasions. No 70014 has transferred to the eastern section where it has been observed on Dover boat trains.

Although Britannias have worked over the mainline to Exeter Central on a couple of occasions authority has not yet been given for them to work beyond this point into north Devon and Cornwall.

No 70009 hauled the up 'Royal Wessex' from Weymouth on 21 August possibly the first appearance of one of this class in the Weymouth area.

RAILWAY TIMES 1951

As mentioned in the 'Brief Encounters' section of this volume, No 70004 sustained a failure on 21 October whilst hauling the 'Golden Arrow' which resulted in all members of the class being temporarily withdrawn for checks. It returned to traffic on 7 December hauling the Victoria – Dover boat train.

Members of the class were to make another appearance on the Southern Region two years later in 1953 following the Crewkerne accident when 35020 *Bibby Line* fractured an axle resulting in the whole Merchant Navy class being withdrawn for checks. The resulting motive power shortage was in part met with the influx of two Britannias from the Midland in the shape of Nos 70030/34 and five from the Western, Nos 70017/23/24/28/29, for a period.

MIDLAND REGION. No 70016 temporarily allocated to Holbeck worked the 3.12pm Leeds City to Morecambe and Carnforth on 2 July.

All of the ER allocated pacifics with the exception of No 70004 which is exhibited at the South Bank, have worked into Manchester London Road often on trains from Birmingham New Street and London Euston.

No 70015 allocated to Camden, worked Euston – Liverpool and return services at the end of July. Special excursions from Euston to Crewe Works were headed by No 70016 on 9 and 23 August with this pacific noted on Kentish Town shed after working a special on 10 August.

No 70015 ostensibly constructed for the WR was still at work between Euston and Liverpool at the end of September.

The first batch of 25 Class 7 pacifics has been completed at Crewe and in September frames were set up for the new Class 6 locomotives to be known as the 'Clans'.

WR pacific No 70015 was on loan to the LMR whilst No 70005 was under test on Rugby testing plant in December, this being its second appearance at the test centre having previously visited from the middle of April until the end of May.

Initial tranche of Standard Class 7 Pacifics delivered in 1951			
No.	Name	Entered service	1951 allocation
70000	*Britannia*	05-01-1951	Stratford
70001	*Lord Hurcomb*	14-02-1951	Stratford
70002	*Geoffrey Chaucer*	06-03-1951	Stratford
70003	*John Bunyan*	01-03-1951	Stratford
70004	*William Shakespeare*	30-03-1951	Stratford / Stewarts Lane
70005	*John Milton*	07-04-1951	Stratford
70006	*Robert Burns*	12-04-1951	Stratford
70007	*Coeur-de-Lion*	25-04-1951	Stratford
70008	*Black Prince*	28-04-1951	Norwich Thorpe
70009	*Alfred the Great*	04-05-1951	Norwich Thorpe / Nine Elms *
70010	*Owen Glendower*	05-05-1951	Norwich Thorpe
70011	*Hotspur*	14-05-1951	Norwich Thorpe
70012	*John of Gaunt*	21-05-1951	Norwich Thorpe
70013	*Oliver Cromwell*	30-05-1951	Norwich Thorpe
70014	*Iron Duke*	02-06-1951	Nine Elms
70015	*Apollo*	12-06-1951	Camden
70016	*Ariel*	18-06-1951	Holbeck (LMR)
70017	*Arrow*	02-06-1951	Old Oak Common
70018	*Flying Dutchman*	25-06-1951	Old Oak Common
70019	*Lightning*	30-06-1951	Newton Abbot
70020	*Mercury*	31-07-1951	Plymouth Laira
70021	*Morning Star*	03-08-1951	Plymouth Laira
70022	*Tornado*	16-08-1951	Old Oak Common
70023	*Venus*	21-08-1951	Plymouth Laira
70024	*Vulcan*	06-10-1951	Plymouth Laira
* On the SR only briefly from w/e 27/6/51 to w/e 20/10/51			

Report on Railway Electrifcation

The 1951 'Electrification of Railways' report, chaired by C M Cock, Chief Mechanical Engineer of the Railway Executive, in many respects reiterated the current position in that London Transport should retain its 630vDC fourth rail system, the Southern Region its 660v DC third rail system and that future main line electrification should take the form of 1500vDC overhead as recommended in the 1927 Pringle Committee's report.

However, this latter decision was promptly overturned as the BTC later inclined towards the 25kvAC system that, at the time of the 1951 report, was being trialled by the SNCF in France with encouraging results and indeed a version was to be tested in the UK on the converted lines serving Lancaster & Morecambe.

This isolated Lancaster area scheme could trace its origins back to the Midland Railway's use of the branch to trial a 6,600v AC overhead electrification at 25 kHz in 1908. In 1951 work was begun to convert the route to 50 kHz as a trial for the upcoming 25kvAC system which would become the new standard for future mainline electrification.

The original electric stock was withdrawn in 1951 to be replaced temporarily by steam hauled services whilst the infrastructure was converted. In 1952 trials began with refurbished surplus three car electric units originally built for the L&NWR 4th rail DC system which had operated the Willesden and Earls Court line until closure to passenger traffic in 1940. These sets continued in use until the branch was eventually abandoned in 1966.

In the body of the report various areas of the UK currently operating with electric traction were highlighted in a series of maps one of which illustrated the extensive 3rd rail system on Tyneside that had been inaugurated by the North Eastern Railway back in 1904.

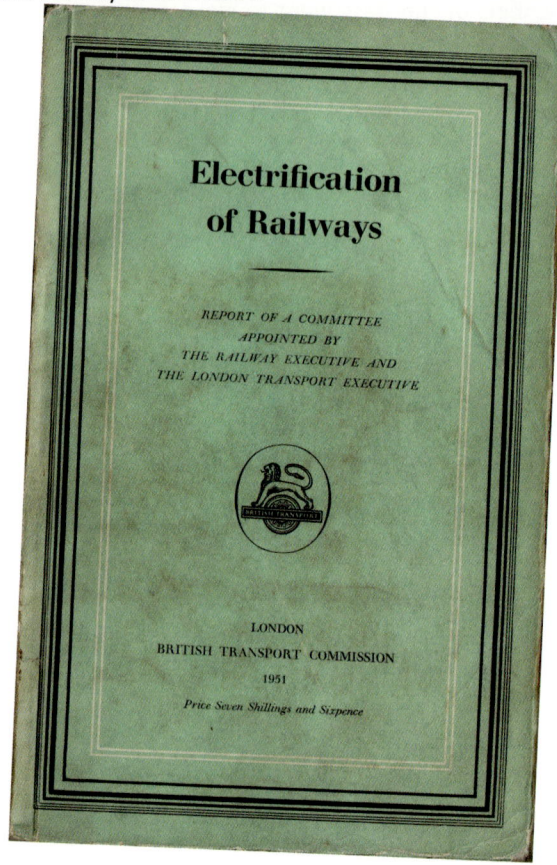

Ex-LMS 1500v DC emu No M28579M is seen operating the Manchester to Altrincham service. These units were later designated as Class 505 and fitted with more modern pantographs but their operation ended in 1971 with conversion of the route to 25kV AC. *OTAHL-ADD094*

RAILWAY TIMES 1951

Electrification on Merseyside encompassed three differing systems, although all were based on current collection using a conductor rail. The lines burrowing under the Mersey were designed as an early electrification scheme to abolish the steam and smoke inherent with steam traction and they utilised a similar four rail system to that espoused by London Underground although at the slightly higher voltage of 650vDC as opposed to London's 630vDC.

The Wirral lines had been electrified by the LMS in 1938 and they used a third rail system although changeover devices were fitted to rolling stock to enable through running. The Liverpool & Southport line took current at 630vDC.

Around Manchester a similar situation of separate schemes and systems was in place with the oldest having its origins in the 1916 Lancashire & Yorkshire Railway scheme to convert the Manchester Victoria to Bury line to electric operation using a 1,200v DC third rail. At Bury the branch to Holcombe Brook had been electrified at 3.5 kV DC overhead by the manufacturers Dick, Kerr & Co Ltd of Preston as a testbed. In 1918 this branch was converted to the Bury line standard of third rail at 1,200v DC.

Like the Bury line the Altrincham line was subjected to heavy electric tram and latterly bus competition and in 1931 was electrified at 1,500v DC overhead, the new national 'standard', by the joint committee of the LNER and LMSR who operated the line. The LNER were arguably the most active of the main line railways in terms of electrification and in the 1930s their attention was directed to their East London lines to Shenfield and to the important freight line across the Pennines between Manchester and Sheffield. Work started before WW2 but was delayed not finally being completed until 1954 with the opening of the newly constructed Woodhead tunnel.

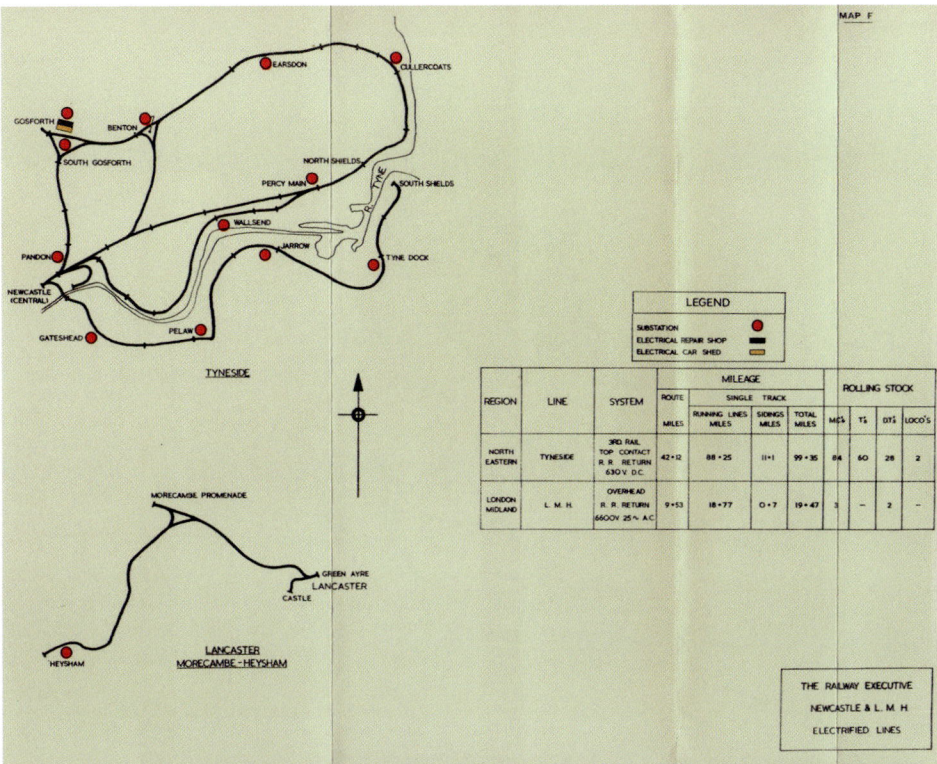

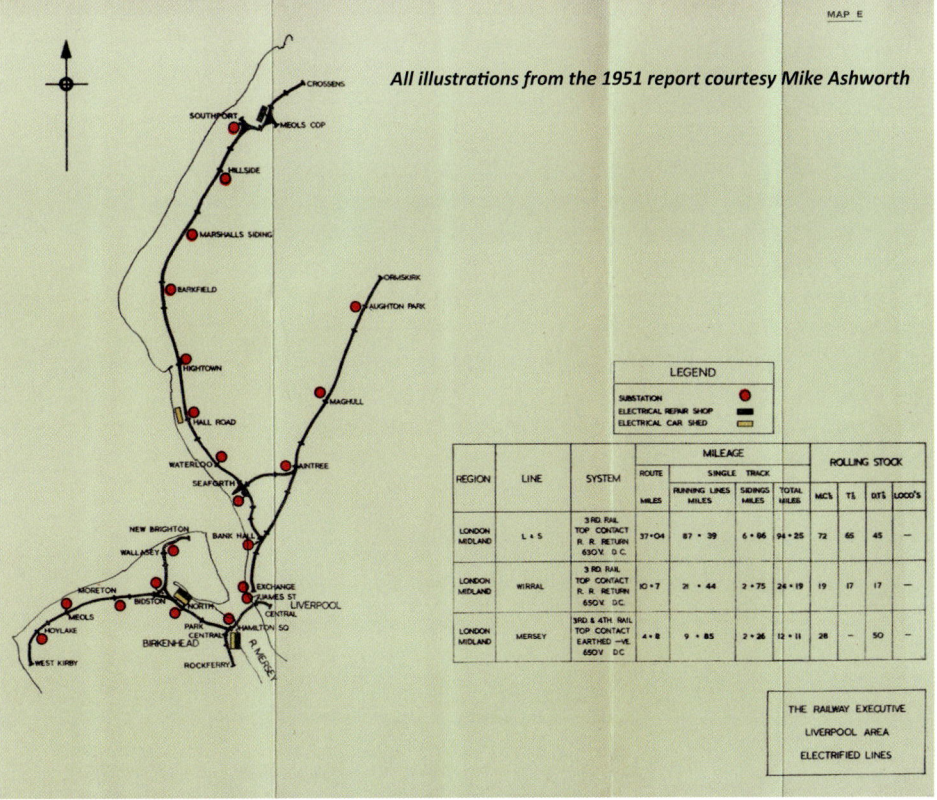

All illustrations from the 1951 report courtesy Mike Ashworth

RAILWAY TIMES 1951

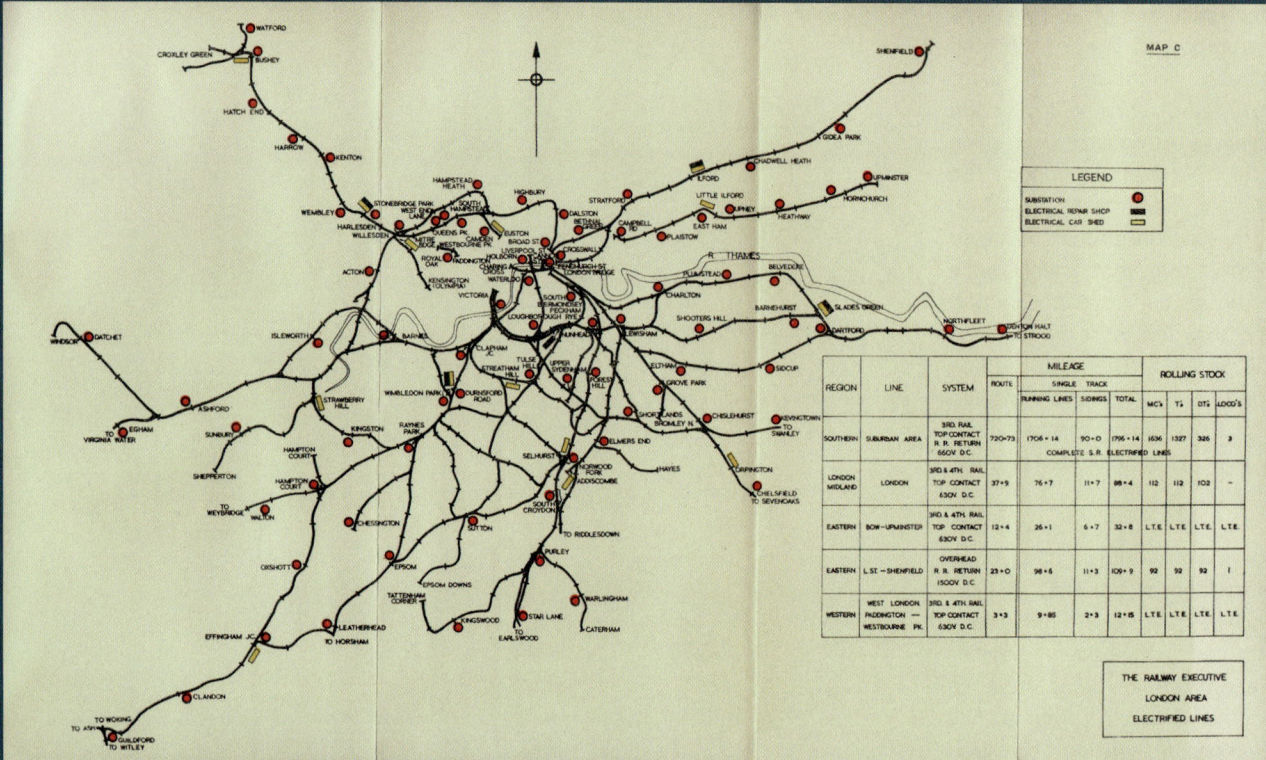

In the capital the majority of the electrified lines were those that the Southern Railway had steadily electrified on the third rail 660v DC system they adopted forming arguably part of the world's largest railway electrification network.

These lines had their origins in pre-WW1 and pre-Grouping companies absorbed into the Southern in 1923. The London & South Western Railway, facing increased tramway competition, had decided on the third rail whilst in south east London the LB&SCR had used a 6,600v AC overhead system similar to that of the Midland Railway's Lancaster system, and this came into use in 1909 as the 'Elevated Electric'. It too saw post-WW1 expansion and even though the Southern decided to adopt the third rail as standard some extensions of overhead continued until 1925. Four years later these extensions were converted to third rail.

North of the river two lines are shown on the map above that were connected with the London Underground thus using their four rail 630v DC system. These are the London Euston - Watford lines that date to WW1 years and that saw joint main line and Bakerloo line services, and the North London lines and the services into the now abandoned Broad Street station. The recently electrified East London line to Shenfield, not long opened in 1949, is also shown.

The final map shows the density of traffic carried over the British Railways system as at 1950/51. Traffic density would be one of the factors that the Committee took into consideration when making decisions upon future electrification schemes and was of course to play a large part in the Beeching study of 1963 where map No. 1 in the report appendix detailed density of passenger traffic/week.

RAILWAY TIMES 1951

Moving forward in time slightly, this is a scene from the opening of the short lived new Woodhead tunnel and the inauguration of the 1,500v DC electrification.
Transport Treasury 2353

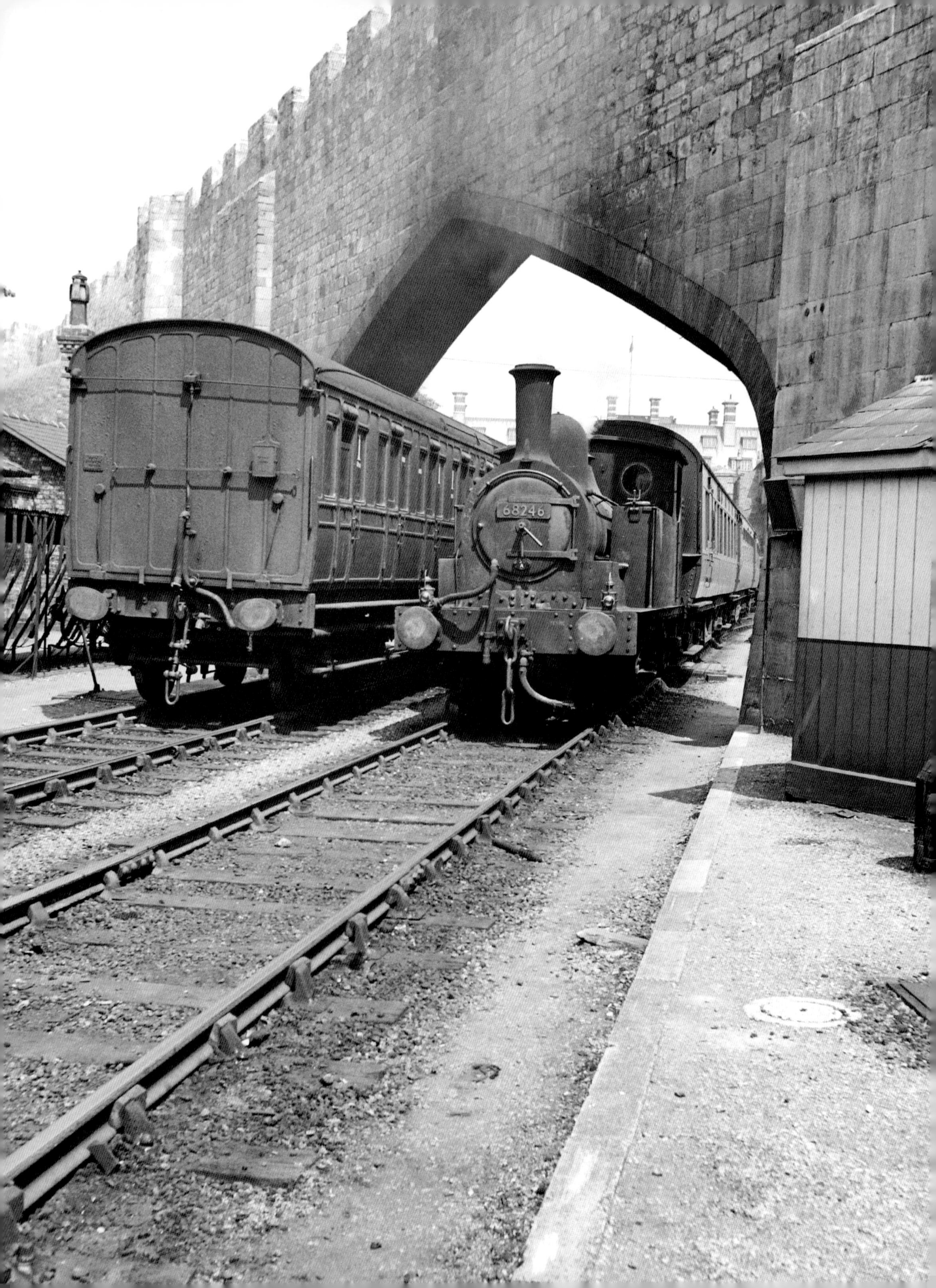
68246

This is York - 1951 Style

Whilst historical York, industrial York and the surrounding countryside form the background to a British Transport film entitled 'This is York' which would be released in 1953 the main action takes place at York station. The hours from dawn to dark on an autumn day are covered in the company of the Station Master who shows something of the planning, hard work and human interest behind the scenes at this key point in the BR network highlighting the workings of the new signalbox which had opened in 1951. At the time this was claimed to be the largest route relay interlocking system in the world covering the work previously undertaken by 70 signalmen with a staff of just 27.

A total of 341 colour light signals and 157 pairs of power operated points replaced semaphores and manually operated points with just one signal box now replacing seven old boxes including Locomotive Yard Box which was believed to have had the largest manual frame at the time in the UK with provision for 295 levers. Under the new regime, 30 miles of track are divided into 316 separate track circuits all of which are connected to a panel in the signalbox showing, by means of lights, current track occupation.

Below: This view showing the inside of the new power signalbox at York dates from 16 November 1951. *Transport Treasury G450*

Opposite: Emerging from York's city walls comes, J71 No 68246 in June 1951. York Station had opened in 1841 and was situated within the city walls. It was designed by George Hudson's (the 'Railway King') associate George Townsend Andrews, the architect who was also responsible for the handsomely proportioned Gothic arch that pierced the ancient city wall. The restricted site made further enlargement of the platforms impossible leading to the opening of the present York through station outside the walls in 1877. The former station retained its tracks which were used for carriage storage until the 1960s when the rails were removed to permit the building of a new office block. *Transport Treasury AF 0246*

Further views of locomotives at York in 1951 appear overleaf.

On York shed on 1 July 1951 is the handsomely proportioned No 50455 one of the Hughes 5P 4-6-0s originating on the Lancashire and Yorkshire railway. This was the last survivor of its class of 60 originals plus 15 rebuilds and was withdrawn in October that year. *Neville Stead NS204033 / Transport Treasury*

Displaying its new ownership and number is J71 No 68246 seen here at York on 1 May 1951. Dating from 1899 this 0-6-0T would put in a creditable 69 years' service before withdrawal in 1958. *Neville Stead NS204508 / Transport Treasury*

Above: 1951 was to see the culling of no less than 18 members of the D20 class . Here one of the survivors No. 62386 is turned on one of York's turntables in June 1951. It would last in traffic until October 1956. *Transport Treasury AF0349*

Right: Class D49 No. 62730 *Berkshire* is seen at York on 21 July 1951. Based at York's North shed from October 1950 until September 1958 it would be withdrawn from nearby Selby shed in November 1958. *Neville Stead NS006748 / Transport Treasury*

Bottom: No visit to York would be complete without a shot of one of Gresley's magnificent pacifics. Here the sleek lines of Class A3 No 60068 *Sir Visto*, which had emerged from Doncaster Works after a light repair two days earlier, are seen to advantage in this view taken at York station on 21 July 1951. It is seen wearing BR blue livery which had been applied in October the previous year. *Neville Stead NS206453 / Transport Treasury*

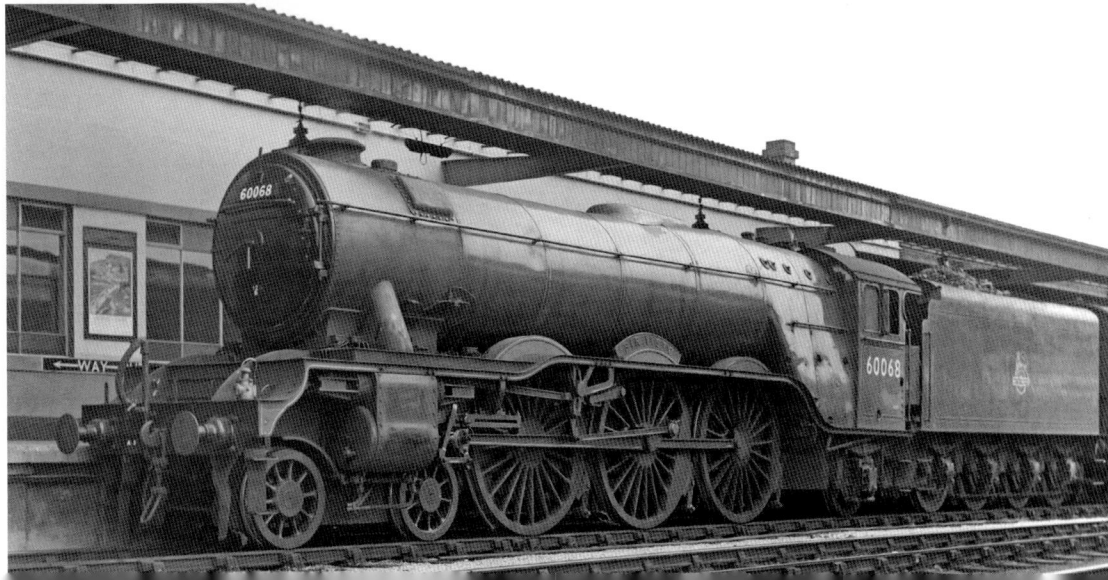

Under the wires at Shenfield comes No. 34039 *Boscastle* on an unrecorded date in 1951. *FH807 / Transport Treasury*

Strange bedfellows at Liverpool Street as Bulleid 4-6-2 No. 34059 *Sir Archibald Sinclair* seen alongside a B12 gets some enquiring glances from passengers and staff alike on an unrecorded date in 1951. As the headboard indicates the pacific was preparing to take out 'The Norfolkman' express. *JCF K2-1 / Transport Treasury*

Anglian 'Spam Cans'

Bulleid's light pacifics had run on Eastern Region metals back in 1948 during the Locomotive Exchanges as detailed in Railway Times No 1.

During the following year another member of the class, No. 34059, made a brief foray into the region during April and May when it was noted working out of Stratford on 2 May with a 9 coach empty stock special to Norwich. It was then tried on 'The Norfolkman' express and over the next four weeks it worked to Norwich and Parkeston Quay as a prelude to the possible transfer of 15 of the class later that year. This trial had been brought about by the problems then being encountered with the region's stock of B1s, B2s and B17s. Although Gerard Francis Gisborne Twisleton-Wykeham-Fiennes better known as Gerry Fiennes, then District Operating Superintendent at Stratford, was mightily impressed with the results the ER did not take up the option of further transfers as the overall trial results were deemed inconsistent caused partly no doubt by the inexperience of crews and depot staff in handling this idiosyncratic machine. All along the ER had held out for a share of the new Britannias so the transfer of Bulleids en masse was never really on the cards.

In 1951 it was reported in the railway press that two of the new Britannia pacifics, Nos 70013 and 70014 were to be temporarily allocated to the SR at Nine Elms in exchange for Bulleid pacifics Nos 34039, 34057 and 34069 which were to be loaned to the ER. Both 34039 and 34057 were allocated to Stratford from May 1951 until March 1952 (34039) and May 1952 (34057) but in the event 34069 was substituted by 34065. They were used primarily on GER line services from London to Cambridge, Norwich and Parkeston Quay as they did not have the tender water scoops required for the longer runs. Following the failure of No 70000 as mentioned in another article in this volume Nos 34076 and 34089 were also allocated to Stratford whilst the Britannias were withdrawn for checks. By May 1952 all examples had been returned to the SR following the reinstatement of the Britannias.

How many railwaymen does it take to turn a Bulleid - seemingly quite a few as five staff struggle to turn No. 34039 *Boscastle* on Cambridge shed's turntable on 12 May 1951.

RAILWAY TIMES 1951

Above: Close up of the contrasting front ends of the two adjacent locomotives. As the B12 has yet to have a smokebox number affixed it is not possible to identify this particular 4-6-0 as the photographer did not record it in his notes. *JCF K2-4 / Transport Treasury*

Middle: "What the +*@! is that thing", might almost be the words of the crew of Britannia No 70006 *Robert Burns* as they look from the cab to observe the strange sight of Bulleid pacific No 34057 *Biggin Hill* seen here on Stratford shed in 1951. *S14 / Transport Treasury*

Bottom: Frustratingly the photographer's notes merely state that this is No 34057 *Biggin Hill* seen at an unnamed junction somewhere in East Anglia. *S49 / Transport Treasury*

The allocations to Stratford were as follows –

34039 *Boscastle* (5/51-1/52)

34057 *Biggin Hill* (5/51-5/52)

34059 *Sir Archibald Sinclair* (4/49-5/49)

34065 *Hurricane* (5/51-5/52)

34076 *41 Squadron* (10/51-4/52)

34089 *602 Squadron* (10/51-12/51).

North Sunderland Succumbs

The North Sunderland Railway opened in 1898 running four miles from Chathill on the East Coast Main Line between Berwick-on-Tweed and Morpeth to Seahouses with one intermediate station at North Sunderland.

Built independently finances were always tight and the 0-6-0ST provided by Manning Wardle was obtained on hire terms. It was named *Bamburgh* after a local settlement which at one time was planned to be joined to Seahouses by an extension which never materialised.

There were initially five trains each way on weekdays with a journey time of 15 minutes. Trains used a bay platform at Chathill and in the 1930s an early diesel shunting locomotive named *Lady Armstrong* was also hired with the line being worked on the 'one engine in steam' principle. *Bamburgh* continued in traffic until 1941 not being scrapped until about 1948.

Motive power until closure was then provided by LNER Class Y7 tank locomotives working with former GER six wheeled coaches. Although traffic declined sharply in the 1930s the line managed to soldier on until its closure on 27 October 1951.

Top and bottom: Two views taken on the same date show Y7 No 68089 awaiting departure time from the terminus at Seahouses. *Neville Stead NS208804A and NS208804B / Transport Treasury*

Above: Former GER six wheeled carriage No. 2 used on the NSR is seen parked at Seahouses on 2 October 1951. *Neville Stead NS208807 / Transport Treasury*

RAILWAY TIMES 1949

The crew proudly pose for the photographer with their Tweedmouth based 0-4-0T mount No 68089 which would be withdrawn in three months time being the penultimate member of its class. However its career was not over for it was sold to a contracting firm who were undertaking works on Morecambe promenade where it could be seen until scrapped in 1955. *Neville Stead NS208804 / Transport Treasury*

And here is No 68089 still wearing its BR livery and number sitting on the promenade at Morecambe. *Neville Stead NS208652A / Transport Treasury*

RAILWAY TIMES 1951

Two coaches and a van are parked by the rudimentary brick and wood station building at Seahouses. Today the station site is a car park. *Neville Stead NS208803A / Transport Treasury*

A member of staff waits expectantly on the board crossing with a barrow whilst the locomotive gets up steam ready to haul its two coaches back to the main line at Chathill. *Neville Stead NS208803 / Transport Treasury*

RAILWAY TIMES 1951

The dimunutive 0-4-0T looks very smart and note the pile of coal on the tank top which you might have thought indicated that these essentially shunting locomotives did not have a coal bunker. However, they did have a small space at the rear of the cab thereby allowing easy access. The base of the bunker opened out into the cab thereby allowing easy access. This coal space was balanced in the opposite tank by the space required to accommodate the reversing gear's movement. As the bunker would typically be covered with coal this gave the false impression that coal was simply piled on top of the left side tank. *Neville Stead NS205004 / Transport Treasury*

RAILWAY TIMES 1951

The headboard 'Farne Islander' was carried on the final day of operation on 21 October 1951 in recognition of the fact that the branchline enabled tourists to reach the harbour at Seahouses and thence take a boat across to the nearby Farne Islands. *Neville Stead NS208806 / Transport Treasury*

Previous motive power was provided in the shape of Manning Wardle 0-6-0T Works No. 1394 named *Bamburgh* seen here at Chathill. *Neville Stead NS208802 / Transport Treasury*

RAILWAY TIMES 1951

Diesel power was to be seen on the branch provided by Armstrong Whitworth 0-4-0DE Works No D25 also seen at Chathill. *Neville Stead NS208801 / Transport Treasury*

Our final view captures No 68089 bowling along the branch which as can be seen was unfenced for much of its length. *Neville Stead NS208805 / Transport Treasury*

Newish (!) 4-car EMU for the Southern Region

Designed by Bulleid the first series of the units, designated 4-EPB, that were constructed were based on Southern Railway designs and were built using standard SR components such as doors and used a standard SR body profile which was mounted on underframes reclaimed from older vehicles the wooden bodies of which had been scrapped. As their designation implies they incorporated electro-pneumatic brakes and they also benefited from buckeye couplings and roller blind headcode displays in place of the stencil holders used previously. Unusually they did not have external doors to the cabs the driver instead accessing via the adjacent guard's compartment.

Unit 5001 was completed at Eastleigh with further examples being constructed there until 1957. In 78 of the total of 213 units built, one car was a former 4-SUB trailer rewired to operate with 4-EPB stock. The formation of each unit was similar to the later 4-SUB units comprising two driving motor brake third open cars with a driving cab, guard's compartment and an eight bay passenger saloon having 82 seats, between which was a trailer third having ten compartments providing 120 seats, with the exception of a couple of cars which only had nine compartments, and a trailer third open having ten seating bays providing 102 seats. The total capacity of the majority of the units was thus 386 seats. Third class was later renamed second class and the car designations changed to driving motor brake second open (DMBSO), trailer second (TS) and trailer second open (TSO).

The first of the units, which had been completed at Eastleigh in October 1951, was reviewed by Robert Riddles member of the Railway Executive for Mechanical & Electrical Engineering at Waterloo on 14 January 1952. The 4 EPBs continued in service, renumbered Class 415 under TOPS, until 1995 becoming the most numerous class of EMUs on the SR following the withdrawal of the 4 SUB units.

Top: Awaiting inspection by R A Riddles of the Railway Executive at Waterloo's platform 11 on 14 January 1952 is new SR EMU No. 5001 the first of the EPBs. *Lens of Sutton, Dennis Culum collection 1223*

Bottom: Seen in revenue earning service is unit No. 5001 leaving Waterloo with the 1:10 Guildford departure yet without any headcode displayed. *Lens of Sutton, Dennis Culum collection 1224*

RAILWAY TIMES 1951

LB&SCR Number	Built	SR Number	Name	BR Number	Withdrawn
37	December 1905	2037	*Selsey Bill*	32037	July 1951
38	December 1905	2038	*Portland Bill*	32038	July 1951
39	January 1906	2039	*La France* (June 1913 - January 1916) *Hartland Point* from January 1926	32039	February 1951
40	February 1906	2040	*St Catherine's Point*	n/a	January 1944
41	February 1906	2041	*Peveril Point*	n/a	March 1944

Last H1 Atlantics withdrawn

The year 1951 saw the last two H1 atlantics withdrawn from service. Designed by Douglas Earle Marsh and constructed by Kitsons in 1905 and 1906 the class soon proved to be successful working London to Brighton express trains including the heavily loaded Pullman services the 'Brighton Limited' and 'The Southern Belle' the latter being described by the LB&SCR as "the most luxurious train in the World".

During the mid 1920s they were gradually replaced on the Brighton run by 'King Arthur' and 'River' class engines but they held their own operating boat trains to Newhaven.

The wartime cessation of the cross-channel ferries left them with little gainful employment and several were put into store or moved to miscellaneous duties with the first members of the class being withdrawn in 1944. No. 39 had been named *La France* in connection with the visit to Portsmouth of the President of France and was subsequently often used for royal duties. The whole class was named, and No 39 renamed 'Hartland Point', during 1925–26 as the Southern Railway's publicity department felt that express passenger locomotives should be named to improve the railway's public image. The names selected were all of coastal headlands in southern England. The H1s were precursors to the later H2 atlantics of which six examples were built with superheaters, larger cylinders and with their boiler pressure reduced to 170 psi.

Opposite top: On 12 July 1947 Newhaven shed had several Atlantics allocated to its strength for working the boat trains from Victoria to Newhaven with two of them seen here. On the right is H1 class No. 2037 *Selsey Bill* whilst on the left is H2 class No. 2423 *The Needles*. Making up the trio of locomotives seen outside the shed is E4 class No 2508. *MPM996-00017 / Transport Treasury*

Opposite bottom: Four years later and heading to the scrapline at Eastleigh on 26 July 1951 is No. 32037 *Selsey Bill* one of the final pair to remain in traffic. In the background is Bulleid pacific No 34056 *Croydon* which was scheduled to enter the Works shortly for a Light Intermediate overhaul. *Neville Stead NS200876 / Transport Treasury*

Above: No 32037 awaits departure from London Bridge with the 5.40 pm service to Tunbridge Wells West on 5 July 1949. *PP523*

RAILWAY TIMES 1951

RAILWAY TIMES 1951

Opposite top: No 32038 *Portland Bill* hurries the 5.40pm from London Bridge to Tunbridge Wells West past New Cross Gate on 21 July 1950. *PP619 / Transport Treasury*

No. 32038 *Portland Bill* negotiates the layout at Clapham Junction with a Brighton to Birmingham service on 9 September 1949. *PP557 / Transport Treasury*

Above: Having already featured in Railway Times No. 1 No. 32039 *Hartland Point* is seen here at Brighton on 31 August 1949 whilst undergoing trials in connection with the Leader experiment with the ungainly sleeve valve modifications. *Neville Stead S200877 / Transport Treasury*

Right: No 2039, still with nameplates attached but now minus rods and ready to be shunted into her final resting place at Eastleigh where she will be dismatled.

RAILWAY TIMES 1951

Top: High Halstow Halt was a stop between Cliffe and Sharnal Street stations on the Hundred of Hoo railway which connected Gravesend with Port Victoria and which was latterly cut back to Grain. Opened in July 1906 it closed to passengers on 4 December 1961 although the track is still used today for freight to Grain. The halt was a typical product of Exmouth Junction concrete works. *Lens of Sutton / Dennis Culum collection 1087*

Bottom: Staying in Kent we have the delightfully named Poison Cross halt situated on the East Kent Railway between Shepherdswell and Richborough. Its first appearance in the timetable was in May 1925 but it was very short lived closing to passengers in November 1928. It had a 50' long corrugated iron faced platform with no shelter provided. Initially it was only open on Saturdays not getting a full service until August 1926 but this proved too excessive with the service cut back to just Wednesdays and Saturdays in November that year until closure. It remained open for freight until January 1950. *Lens of Sutton / Dennis Culum collection 0913*

Let's Halt Awhile

Ashley Courtenay's famous guidebook 'Let's halt awhile in Britain' ran to many editions and was to be found on the book shelves of many a home during the 1950s and 1960s. In this feature we highlight a selection of lines which housed some of the many small, often very basic, rural halts still to be found on the BR network in 1951 - if you knew where to look.

Top: An old coach body adorns Cutler's Green halt on the Elsenham and Thaxted light railway. Opened in April 1913 it closed in September 1952. The only access to this halt was via a footpath. *On-Line Transport Archive 3450*

Bottom: Also on the Thaxted line was Sibleys which, as the nameboard proudly proclaims, was 'For Chickney & Broxted'. Rather unusually trains shunted here using a towrope. There was a goods loop and two short sidings at either end of the loop and shunting was performed by attaching a towrope to wagons in the loop with the locomotive manoeuvring them from its position on the mainline. *Neville Stead NS207361 / Transport Treasury*

RAILWAY TIMES 1951

Top left: Another stop on the Thaxted route, which seemed to specialise in bucolic wayside halts, was Henham halt consisting of a 100 foot long raised clinker and ash platform on the north side of the line, an old 4-wheel coach body as a shelter and rather unusually for such an insignificant stop a brick built gentlemen's toilet. *Neville Stead NS207360 / Transport Treasury*

Top Right: The first stop on the branch after leaving the junction at Elsenham was Mill Road Halt built in 1922 to the south of the village. In line with keeping costs to a minimum this halt was very basic with a short 25 foot platform just 1 foot in height with a clinker surface and timber facing. A small wooden hut acted as a waiting shelter with a name board and two lamps completing the facilities. *Neville Stead NS207359 / Transport Treasury*

Bottom left: Inworth halt, sporting an interesting waiting room still adorned with an LNER poster board in this 1951 view, was on the Kelvedon & Tollesbury Railway (see further article on this line in this volume, pages 9-12) serving the village of Inworth in Essex. Opening in 1904 it closed, along with the rest of the line, on 7 May 1951. Passenger accommodation at stations was to say the least eccentric, being an old bus body at Feering whilst at Inworth, Tolleshunt Knights and Tolleshunt d'Arcy the bodies of old four-wheeled coaches were used. *Neville Stead NS207369 / Transport Treasury*

Bottom right: Here is that aforementioned bus body doing duty at Feering halt. This is believed to be part of the body from an AEC 'B' type vehicle commonly known as an 'Ole Bill' which were used by London Transport from 1911 and which saw service as troop transport in WW1. Its nickname derived from the famous Bruce Bairnsfather cartoon featuring two Tommies under fire in a shell hole with Bill saying to his companion, "If you knows of a better 'ole, Go to it!" *Neville Stead NS207368 / Transport Treasury*

Opposite top: Grimethorpe Halt was on the Dearne Valley Railway on the line between Wakefield and Edlington. The station opened on 3 June 1912 closing on 10 September 1951. This view was taken on 18 August the month before closure and shows Ivatt tank No 41250 with a handful of passengers occupying the single coach No 3446 which was a pull-push conversion of older LMS gangwayed stock as witness the corridor connection still in situ. *Neville Stead NS204764 / Transport Treasury*

Opposite bottom: This line was built in 1886 by the Taff Vale Railway to serve the Lady Windsor Colliery near Robertstown. Initially the line was used only for coal and freight traffic and although a passenger service to Abercynon followed in 1890 it was not until the introduction of steam railmotors, which occurred in 1904, and diversion into Pontypridd that the service became successful. The route was steeply graded with a 1 in 47 incline up to Ynysybwl. Closure to passengers came in 1953 but the line remained in place for use by the Lady Windsor Colliery until its closure in 1988. Here 64xx class pannier tank No 6411 has arrived with its one coach train at Old Ynysybwl halt on 11 September 1951. *BMS BM0234*

RAILWAY TIMES 1951

North of the Border in '51

This page, top: D40 4-4-0 No 62269 is seen at Boat of Garten's bay platform with a service for Craigellachie and Keith on an unrecorded date in 1951. Dating from 1913 this example was allocated to Keith depot and remained in service until the autumn of 1955. One example has been preserved in the shape of *Gordon Highlander* GNoSR No. 49. *Neville Stead NS207923 / Transport Treasury*

This page bottom: Banchory on the branch from Aberdeen to Ballater sees the arrival of B12 No 61552 on 7 July 1951 having deposited a goodly crowd of passengers. Inroads were made into this class of 4-6-0 in the early 1950s with this particular locomotive being withdrawn from service from Kittybrewster shed the following year. *Neville Stead NS206202 / Transport Treasury*

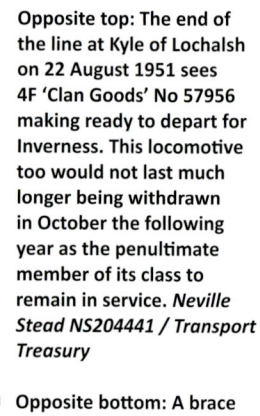

Opposite top: The end of the line at Kyle of Lochalsh on 22 August 1951 sees 4F 'Clan Goods' No 57956 making ready to depart for Inverness. This locomotive too would not last much longer being withdrawn in October the following year as the penultimate member of its class to remain in service. *Neville Stead NS204441 / Transport Treasury*

Opposite bottom: A brace of Class Y9 0-4-0STs are seen on Edinburgh's St. Margaret's shed on 6 October 1951. No 68098 had just been renumbered that month replacing its previously held LNER number of 8098. One member of the 38 strong class has been preserved by the Scottish Railway Preservation Society in the shape of the former No 68095. *Neville Stead NS202883A / Transport Treasury*

61

RAILWAY TIMES 1951

Delivered from the Birmingham RC&W Co. in 1951, *Pegasus* is seen in 'as new' condition at Clapham Junction early that year. It would be given the name "Trianon Bar" and ran regularly in the 'Golden Arrow' formation.
Lens of Sutton, Dennis Culum collection 0996

Golden Arrow Pullmans & the 'Trianon' Bar(s)

Pullman car *Pegasus* illustrated opposite was one of the new Pullman cars built in 1951 as part of the Festival of Britain celebrations. This coach had the bar area named *The Trianon Bar* as was traditional for these vehicles. These 1951 cars were distinctive in that they were built with rectangular rather than oval lavatory windows and whilst the interiors no longer had marquetry panelling each did differ in the type of panelling used although the cars did become perhaps best well known for their mosaic toilet floors. *Pegasus* subsequently went on to serve in the night sleeper from Euston with her bar renamed "The Nightcap Bar".

The 'Golden Arrow' saw the introduction of seven new cars and three refurbished cars to the train's formation in 1951 with a further three new cars being introduced in 1952. Originally the new cars were earmarked for service on the LNER and work commenced on them in 1938 but the outbreak of the Second World War brought construction to a halt. All the cars retained Gresley bogies thus identifying their LNER heritage. Details of the new cars shown below.

A word of explanation may be in order here about the vehicles that have served as the *Trianon Bar*. In the space of 17 years there were no less than three vehicles so named with the first being a conversion undertaken in 1946 at Preston Park from an eight wheeled 1928 kitchen car built to schedule No 250 and originally named *Diamond*.

The second was 12 wheeled third class kitchen car No 5 built to schedule No 68 in 1917 at Longhedge and converted at Preston Park early in 1946. This latter vehicle was a showpiece as the interior was entirely fitted out in 'Wareite' laminated plastic in grey, pink and cream. Even the curtains were plastic in a rather fetching shade of pink! There were no seats with all the space being devoted to the bar. It entered service upon the post war re-inauguration of the 'Golden Arrow' on 15 April 1946. Developing a hot axle box on its inaugural trip it was replaced by the former *Diamond* Car No 5 and was then briefly renamed the *New Century Bar* and placed on the London Victoria - Dover Ostend service.

However as this car was always intended for the 'Golden Arrow' in July 1946 it was taken off the Ostend train, renamed *Trianon Bar,* and put back on the 'Golden Arrow' service. *Diamond* then returned to the Ostend service where it remained until 1952. It was later refurbished and put into service on the South Wales Pullman in 1955 under the name *Diamond : Daffodil Bar* presumably as a nod to the Welsh national flower.

With the introduction in 1951 of a new 'Golden Arrow' set the *Trianon Bar* was replaced with *Pegasus* built to schedule No 310 which contained the bar plus 14 first class supplement seats.

The twelve wheeled car reverted to its original number 5 as a buffet car and ended its days as a camping coach. The final 'Golden Arrow' service operated on Saturday 30 September 1972 and consequently all Pullman cars remaining on the service except *Pegasus* were withdrawn.

Pegasus was then used as a 'Nightcap Bar' on West Coast Main Line Sleeping Car Trains, a role it continued to perform for several subsequent years. By the way the word 'Trianon' means a small elegant villa the most famous of which were the Grand and Petit Trianon at the Palace of Versailles commissioned by Louis XV1. (Ed: All of the 1951 cars are still with us today, with the exception of *Hercules* which was broken up at Clapham Junction. *Pegasus*, withdrawn in 1972, was used by HM The Queen on 9 September 2015 for the opening of the Borders Railway in Scotland and today it can be found at Crewe Heritage Centre).

Cygnus	First Parlour	26 seats
Hercules	First Parlour	26 seats
Perseus	First Parlour	26 seats
Pegasus the Trianon Bar	First Parlour / Bar	14 seats
Aquila	First Kitchen	22 seats
Orion	First Kitchen	22 seats
Carina	First Kitchen	22 seats
In addition to the new cars, three older cars were rebuiilt and refubished		
Minerva	First Brake Parlour	26 seats
Car No 35	Second Parlour	42 seats
Car No 208	Second Brake Parlour	36 seats

Opposite bottom: With Golden Arrow insignia on the side and named 'The Trianon Bar' Pullman car *Pegasus* is seen at Victoria. Whilst the new square windows in the toilet compartment are evident in this view it may also be seen that the windows on the doors retained their traditional oval shape.

RAILWAY TIMES 1951

RAILWAY TIMES 1951

Opposite top: In splendid isolation *Pegasus* stands outside the Preston Park Pullman Works in Brighton.

Opposite bottom: The interior of *Pegasus* reveals, behind the barman, one of the pair of aluminium plaques used to decorate the interior, this one depicting the Eiffel Tower apparently the target of a 'Golden Arrow' shot by Cupid.

This page, top: The rake of 'Golden Arrow' stock including The Trianon Bar is shunted at Stewarts Lane carriage sidings. All passengers regardless of their class of ticket were entitled to use the *Trianon Bar*.

This page, bottom: The earlier twelve wheeled wooden bodied Pullman named *Trianon* and carrying the name *Trianon Bar* is seen by the chalk cutting at Preston Park. The oval lavatory windows are evident in this earlier design.

RAILWAY TIMES 1951

It looks like cocktail hour in this view of the interior of Pullman Car *Diamond* which rather overdoes the flower motif although this does perhaps help to explain its later title as the *Daffodil Bar*.

Opposite top: With the station clock showing 12.15 pm the queues for Saturday holiday services line the concourse waiting behind the numerous portable boards displaying the various destinations of the many holiday trains. Motive power in view includes a 'Lord Nelson' on the far right in front of which a Bulleid pacific simmers at the buffer stops with a 4 COR 'Nelson' EMU at platform 12. *Lens of Sutton Dennis Culum collection 1056*

Opposite bottom: Looking further to the left but on a different occasion in the same year is a rare visitor, more usually seen at Victoria deployed on 'Brighton Belle' services, in the shape of one set of the 5 BEL EMUs. Although the exact date of this view in 1951 went unrecorded by the photographer it is almost certainly 20 July 1951 the occasion when the unit seen here, No 3052, conveyed HRH Princess Elizabeth, then Duchess of Edinburgh, to Portsmouth & Southsea in Pullman car *Audrey*. A Pathe Newsreel of the visit is available on Youtube with an opening shot of the Pullman train at Portsmouth. Whilst in the town she opened a new United Services club and the restored Connaught drill hall. These 5 BEL units were used from time to time on specials, including those carrying the Royals, both on and off the Brighton line. Alongside at platform 8 a Bulleid pacific rests at the buffer stops. *Lens of Sutton Dennis Culum collection 1043*

From a Waterloo Eyrie

High up among the roof girders at Waterloo were a number of offices including room 119. From this vantage point our photographer took these two views in 1951.

CRANE TANK BOWS OUT AT BOW

This view of LMS Class 0F Crane Locomotive No 27217 was taken at Bow Works in 1938. Originally built as an 0-4-0ST for the North and South Western Junction Railway by Sharp Stewart & Co. in 1858, this locomotive was later converted to an 0-4-2CT with a steam crane carried over a new trailing truck. Inherited by the LMS from the North London Railway it has borne a variety of numbers in its long career being 37, 29 and 29A on the NLR, 2896 on the L&NWR and finally 7217 and later 27217 on the LMS. It worked exclusively at Bow Works as a service locomotive remaining operational throughout the war years and at Nationalisation becoming BR's oldest serving locomotive. The crane had a 3 ton capacity with coal bunkers on either side and the front buffer beam was a solid iron casting as an aid in balancing the locomotive. It was withdrawn from Bow in February 1951 numbered 58865 by BR after 93 years of service but was not scrapped until October 1952 at Derby. *27217 bow works*

Rail Transport at the Festival of Britain

Transport featured heavily in the Festival with a whole pavilion being devoted to the subject covering various aspects including rail, air, sea and road. Communications was also included in this pavilion featuring the latest developments in radio, television and postal services.

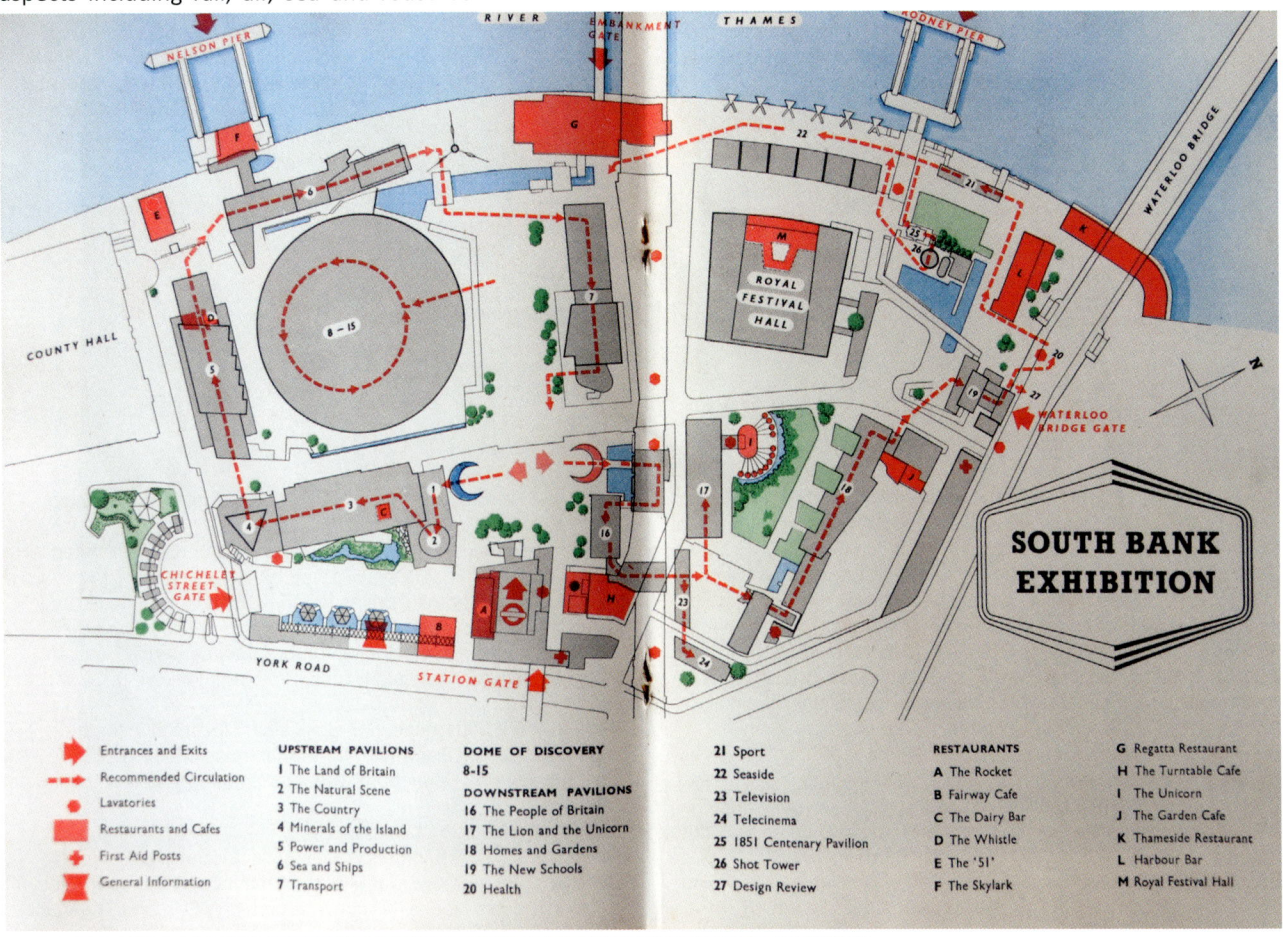

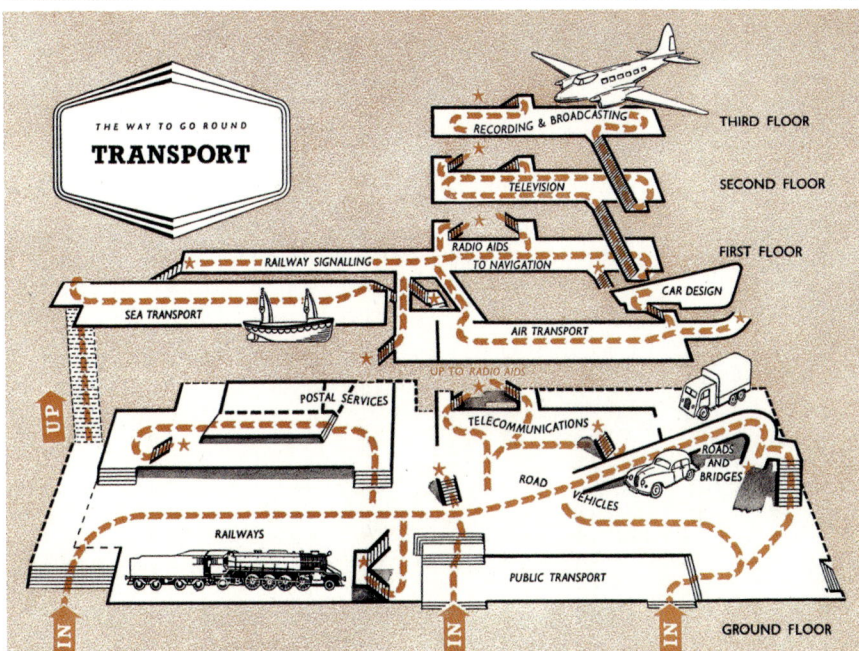

Plan of the Transport Pavilion with the Indian Railways locomotive outside as it was too large at 120 tons to be accommodated inside. It was transported from Scotland by sea to London and then by road to the site where it was off loaded still on its trailer on to a platform and rolled sideways into place outside the pavilion.

RAILWAY TIMES 1951

Dodging the spray from the fountain, a couple of young girls enjoy a snack looking out over the area in front of the Transport Pavilion. Next to the suspended lifeboat rests 'WG' 2-8-2 locomotive No 8350 built by North British in Glasgow to the 5ft. 6 in. gauge of Indian Railways. In front of it can be seen a broad gauge 8 foot wheelset which legend has it originally belonged to the GWR's 4-2-2 *Lord of the Isles*.
R E Vincent REV 58A 1-1, Transport Treasury

SOUTH BANK TRANSPORT & COMMUNICATIONS EXHIBITS.
A501 Three types of early timber rail (1738); The Science Museum, South Kensington, London S.W.7
A502 Cast iron plate rails from the Pickering Lime Quarry Tramway (1775); The Railway Executive (British Railways).
A503 Fish-bellied edge-rail from the Loughborough and Nanpanton Railway (1789); The Railway Executive (British Railways).
A504 Original Mansfield & Pinxton rail with stone block supports (c1818) ; Borough of Mansfield Museum & Art Gallery, Mansfield,
A505 Original drawing of Lawson Main wagon road; City of Birmingham Public Library, Birmingham.
A506 Length of wrought iron rail of type used in 1808; The Railway Executive (British Railways).
A507 Fish-bellied malleable iron T-section rail, Stockton & Darlington Railway (1825); The Railway Executive (British Railways).
A508 Bull-head type of rail with lower flange smaller than the head; The Railway Executive (British Railways).
A509 Flat-bottom rail with joint chairs, York and North Midland Railway (1839), originally invented by Colonel Stevens; The Railway Executive (British Railways).
A510 Double-headed rail jointed by fishplates shown with chair (1835); The Science Museum, South Kensington, London S.W.7.
A511 Fishplate joint (W. Bridges Adams and R. Richardson) (1847); The Science Museum, South Kensington.
A512 Bessemer steel rail, Midland Railway (1870); The Railway Executive (British Railways).
A513 Stone block sleeper with short section of rail; The Railway Executive (British Railways).
A524 *Agenoria* locomotive (1829); The Science Museum, South Kensington, London S.W.7. and the York Railway Museum, York.
A525 Frame and rollers for A524; Hoffman Manufacturing Co. Ltd., 8 Bedford Square, London W.C.1.
A526 Original early drawings of export locomotives; The Vulcan Foundry Ltd., Newton-le-Willows, Lancs.
A527 Chain and sprocket drives for A524; Renold and Coventry Chain Co. Ltd., Coventry, Warwicks.
A528 Switchgear and electric motor for A524; Laurence Scott and Electromotors Ltd., Manfield House, Southampton Street, Strand, London W.C.2.
A529 Worm reduction gear unit for A524; Highfield Gears & Engineering Co. Ltd., 119 Victoria Street, London S.W.1.
A539 Worm reduction Rear unit for A54; Highfield Gears & Engineering Co. Ltd.

A540 2-2-2 Buddicom Locomotive (1843); Societe Nationale des Chemins de Fer Francais, Paris 9.
A541 Frame and rollers for A540; British Timken Ltd., 22 Bruton Street, London W.1.
A542 Chain and sprocket drives for A540; Renold and Coventry Chain Co. Ltd., Coventry, Warwicks.
A543 Switchgear and electric motor for A540; Laurence Scott and Electromotors Ltd., Manfield House, Southampton Street, Strand, London W.C.2.
A544 2-8-2 5-ft 6-in-gauge WG-class steam locomotive for the Indian Government Railways; North British Locomotive Co. Ltd., 110 Flemington Street, Springburn, Glasgow; by permission of the Indian Government.
A552 Water pick-up installation.
A560 660 b.h.p. 3-ft 6-in-gauge Diesel-electric locomotive, for the Tasmanian Government; wheel arrangement Bo-Bo; English Electric Co. Ltd., Queen's House, Kingsway, London W.C.2, and The Vulcan Foundry, Newton-le-Willows, Lancs; by permission of the Tasmanian Government.
A561 3-ft 6-in Colonial-gauge track with 60-lb per yard (29.76 kgs per metre) revised British Standard flat-bottomed rails on steel sleepers with various fastenings; The Permanent Way and Accessory Manufacturers of Great Britain.
A562 4-ft 8-in-gauge 300/330 bhp. 0-6-0 heavy duty Diesel-mechanical shunting locomotive (Stephenson-Crossley); Robert Stephenson & Hawthorns Ltd., Darlington, Co. Durham.
A563 Standard 4-ft 8-in-gauge track with 95-lb per yard (47.13-kgs per metre) bull-headed rails on concrete sleepers with cast -iron chairs and plate sleepers; The Permanent Way and Accessory Manufacturers of Great Britian.
A564 *Planet* 7-ton 22-in-gauge 37 bhp. Diesel-mechanical locomotive with 2-speed gear box and air brakes; F. C. Hibberd & Co. Ltd., 56 Victoria Street, London S.W.1. and Arthur Guinness, Son & Co. Ltd., Dublin and London.
A565 Track for A564; Rail Makers' Association, 94 Petty France, London S.W.I.
A566 10-ton flame-proof Diesel mines locomotive, 75 h.p., 500-mm gauge, oil-operated gearbox, 4-speed constant mesh; Ruston & Hornsby. Ltd., Lincoln.
A567 Track for A566; Rail Makers' Association, 94 Petty France, London S.W.1.
A568 Petrol-driven railway motor inspection car, metre-gauge, 85 b.h.p.,; D. Wickham & Co. Ltd., Ware, Herts.

RAILWAY TIMES 1951

A569 Track for A568; Rail Makers' Association, 94 Petty France, London S.W.1.
A570 10-mm scale model of 0-4-0 steam shunting locomotive; W. G. Bagnall & Son Ltd., Castle Engine Works, Stafford.
A571 10-mm scale model of 'Uniline' transport system locomotive; J. Brockhouse & Co. Ltd., Victoria Works, Hill Top, West Bromwich, Staffs.
A572 10-mm scale model of 1,000 bhp main line Diesel-electric locomotive; The Brush Electrical Engineering Co. Ltd., Loughborough, Leics.
A573 10-mm scale model of 3,000 hp main line electric locomotive and overhead transmission cables for E.F.S.J. (late Sao Paulo Railway), Brazil; English Electric Co. Ltd., Queen's House, Kingsway, London W.C.2., The Vulcan Foundry Ltd., Newton-le-Willows, Lancs., and British Insulated Callender's Cables Ltd., Prescot, Lancs.
A574 10-mm scale model of 'Planet' 41/50 hp Diesel-mechanical locomotive; F. C. Hibberd & Co. Ltd., Park Royal, London N.W.10.
A575 10-mm scale model of 500 hp Diesel shunter for the Peruvian Corporation; Hunslet Engine Co. Ltd, Jack Lane, Leeds 10, Yorks.
A576 10-mm scale model of 1,070 hp 3,000 volts D.C. electric locomotive for the Rede Mineira de Viacao; Metropolitan-Vickers Electrical Co. Ltd., Trafford Park, Manchester 17.
A577 10-mm scale model of experimental 200 hp gas turbine-electric locomotive for British Railways (Western Region); Metropolitan-Vickers Electrical Co. Ltd.
A578 10-mm scale model of 45-ton Diesel-mechanical shunting locomotive; Ruston & Hornsby Ltd., Lincoln.
A579 10-mm scale model of 4-8-2 19D-class steam locomotive for South African Railways; Robert Stephenson & Hawthorns Ltd., Darlington, Co. Durham.
A580 10-mm scale model of 2-8-0 'Liberation' class steam goods locomotive for U.N.R.R.A; The Vulcan Foundry Ltd., Newton-le-Willows, Lancs.
A589 10-mm scale model of 1,600 hp main line Diesel-electric locomotive for Egyptian State Railways; English Electric Co. Ltd., Queen's House, Kingsway, London W.C.2 and The Vulcan Foundry Ltd., Newton-le-Willows, Lancs.
A590 Brunel's G.W.R. track; The Railway Executive (British Railways).
A59I Pair of early G.W.R. wrought iron hand-forged 8-ft diameter driving wheels with crank axle; The Railway Executive (British Railways).
A592 Standard 5-ft 6-in-gauge track, 90-lb per yard (44.64-kgs per metre), on steel sleepers with spring steel loose jaws and taper keys; The Permanent Way and Accessory Manufacturers of Great Britain.
A593 Standard 4-ft 8-in-gauge track with new standard 109-lb per yard (54.1-kgs per metre) flat-bottomed rails, timber sleepers, steel and cast iron base plates and various types of fastenings; The 'Permanent Way and Accessory Manufacturers of Great Britain.
A594 Metre-gauge 80-lb (39.68-kgs per metre) British Standard flat-bottomed rails, steel sleepers with various fastenings for British Colonies and India; The Permanent Way and Accessory Manufacturers of Great Britain.
A595 Model showing working of A597; - George Turton, Platts & Co. Ltd., Meadow Hall Road, Wincobank, Sheffield, Yorks.
A596 2-ft 6-in-gauge mines track, 35-lb per yard (17.36-kgs per metre), flat-bottomed rails, timber sleepers, dogspikes and steel spring spike fastenings; The Permanent Way and Accessory Manufacturers of Great Britain.
A597 Pneumatic buffers for railway termini stops, employing air in compression for high energy absorption and low recoil energy; George Turton, Plans & Co. Ltd., Meadow Hall Road, Wincobank, Sheffield, Yorks.
A598 Hydraulic track jack; Kenitra Co. Ltd., 48-50 St. Mary Axe, London E.C.3.
AS99 Railway permanent way track gauge; Henry Williams Ltd., Railway Appliance Works, Darlington, Co. Durham.
A600 Railway switch lever; Henry Williams Ltd.
A601 Fully-welded rail-joint cut away to show weld; A.I. Electric Welding Machines Ltd., Rose Street, Inverness.

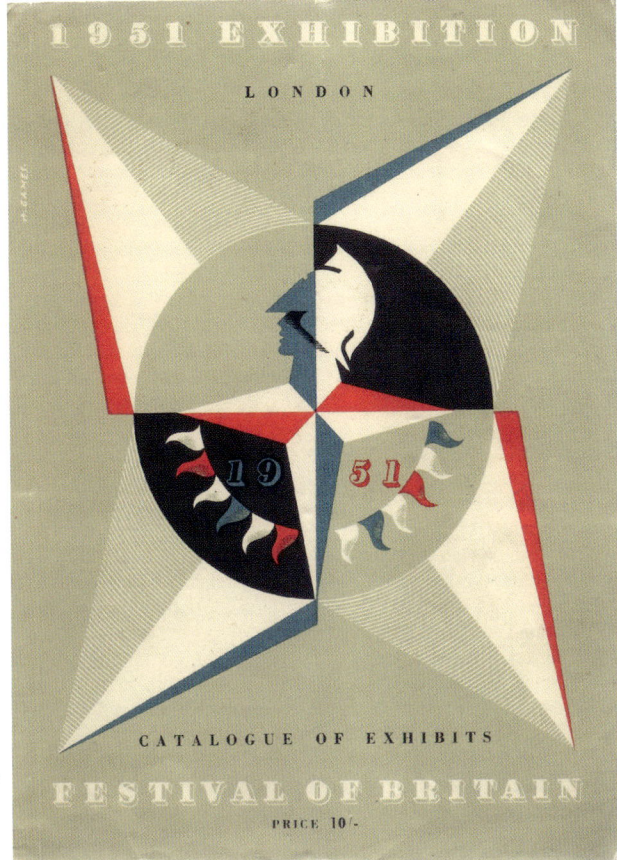

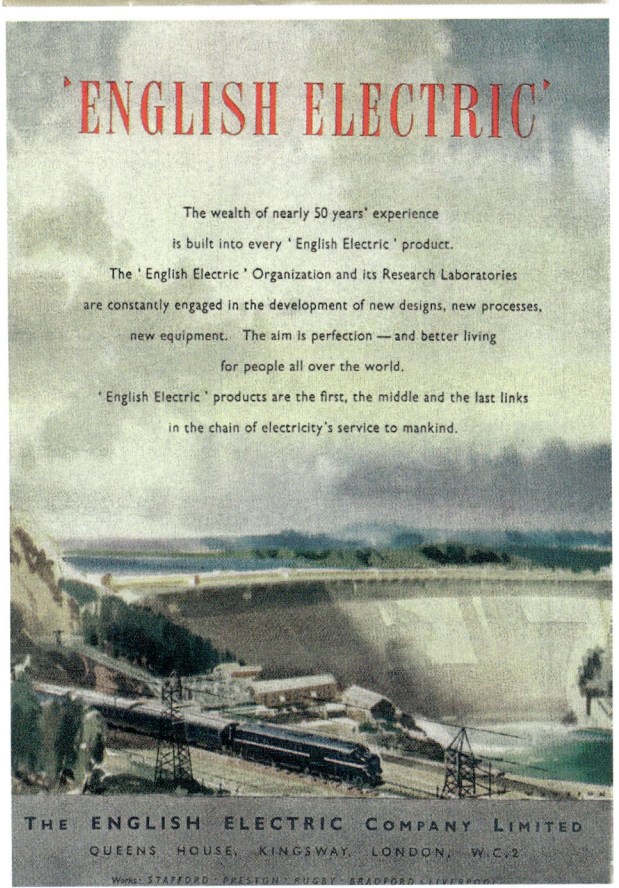

RAILWAY TIMES 1951

A603 Copies of original drawings of early railway carriages.
A610 Original drawing of ceremonial carriage for Viceroy of Egypt (1858); Metropolitan-Cammell Carriage & Wagon Co. Ltd., Saltley, Birmingham 8.
A61I Light-weight non-driving motor car embodying the latest developments of aluminium alloy construction ; designed by the Metropolitan-Cammell Carriage & Wagon Co. Ltd., to the specification of the London Transport Executive ; Metropolitan-Cammell Carriage & Wagon Co. Ltd.
A625 Rail de-icing bath; London Transport Executive.
A626 10-mm scale model of bogie tank wagon for Malayan State Railway; Hurst Nelson & Co. Ltd., Motherwell, Lanark.
A627 10-mm scale model of passenger rail car for the Jamaican Government Railway; D. Wickham & Co. Ltd., Ware, Herts.
A628 10-mm scale model of 50-ton air-operated dump car; Head Wrightson & Co. Ltd., Teesdale Iron Works, Thornaby-on-Tees, Yorks.
A650 'Gloster' heat-treated cast steel bogie, 3-piece assembly; Gloucester Railway Carriage & Wagon Co. Ltd., Bristol Road, Gloucester.
A700 The building of the railway.
A706 Model of modern track-laying machine; The Railway Executive (British Railways).
A720 Model of 'Rurik' ice locomotive exported to Russia (1861); The Science Museum, South Kensington, London S.W.7 and North British Locomotive Co. Ltd., 110 Flemington Street, Springburn, Glasgow.
A721 Britain as a world railway builder.
A765 Model of Westinghouse compressed air automatic type brake; Westinghouse Brake & Signal Co. Ltd., Chippenham, Wilts.
A766 Sectioned model of 'SJ' vacuum ejector; Gresham & Craven Ltd., Ordsall Lane, Salford 5. Lancs.
A767 Caprotti valve gear cambox; Associated Locomotive Equipment Co. Ltd., St. James's Square, London S.W.I.
A768 Rotary campoppet valve gear cambox ; Associated Locomotive Equipment Co. Ltd.
A769 Sectionalised driver's brake valve ; Billington and Newton Ltd., Longport, Stoke-on-Trent, Staffs.
A770 The Thames Tunnel.
A775 The Metropolitan Railway and its construction.
A776 The Tower Subway.
A778 Hydraulic shield pump, part of A780; Hayward Tyler & Co. Ltd., Luton, Beds.
A779 Belt conveyor, part of A780; The Mining Engineering Co. Ltd., Meco Works, Worcester.
A780 Rotary tunnelling shield exerting 400 tons pressure; owners Charles Brand & Son Ltd., 25 Charles II Street, London S.W.1 ; manufacturers: Markham & Co. Ltd., Chesterfield, Derbys.
A78I Cast iron tunnel lining, part of A780; The Stanton Iron Works Co. Ltd., Nottingham.
A782 The London Underground.
A800 The Travelling Post Office.
A807 Original car of the Pneumatic Despatch Co. Ltd. (1861); Borough of Tottenham Museum, London N.17.
A811 Full size working length of Post Office (London) Railway; U.K. Post Office.
A851 Historic and contemporary railway tickets; The Railway Executive (British Railways).
A852 Original Edmondson ticket-dating press (1840); The Railway Executive (British Railways).
A855 'Rapid Printer' ticket-printing and issuing machine; Westinghouse Garrard Ticket Machines Ltd., York Way, King's Cross, London N.I.

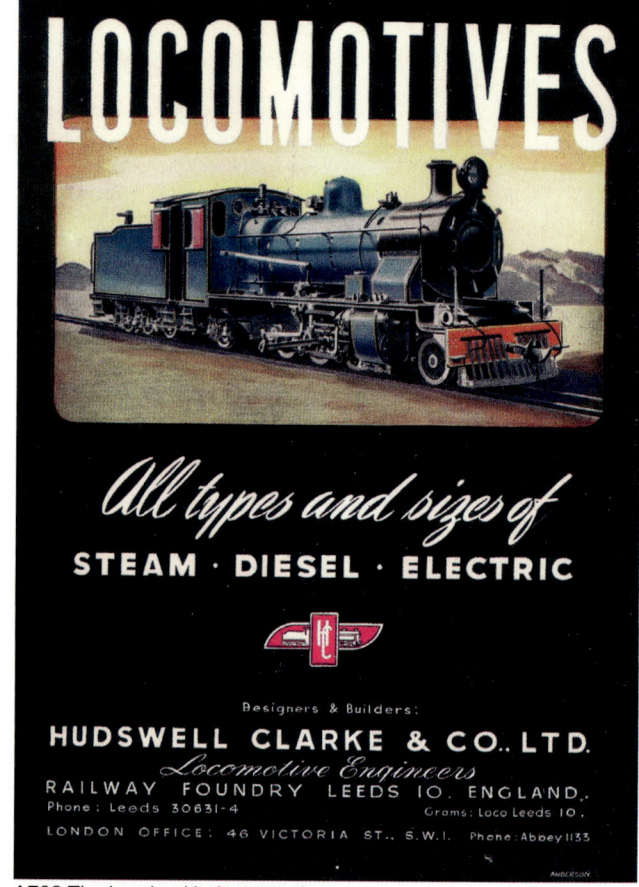

Above: Locomotive builders such as English Electric and Hudswell Clark took the opportunity to advertise their wares in the catalogue.

Left: A locomotive, which upon magnification looks to be a Schools class 4-4-0, is glimpsed on Hungerford railway bridge showing the proximity of BR's running lines into Charing Cross to the eastern boundary of the Festival site. *R E Vincent REV 58A 1-6, Transport Treasury*

RAILWAY TIMES 1951

A856 Electrically-operated railway ticket-dating machine; Bell Punch Co. Ltd., Uxbridge, Middx.
A857 Coin-operated ticket-printing/issuing and change-giving machine; Brecknell, Munro & Rogers Ltd., Pennywell Road, Bristol 5.
A858 Germ-proof booking office window; Hygia-phone (England) Ltd., 140-142 Long Acre, London W.C.2.
A859 Contemporary wall-type coin-operated vending machine for sweetmeats, etc.; British Automatic Co. Ltd., 14 Appold Street, London E.C.2, and Brecknell, Munro & Rogers Ltd., Bristol 5.
A860 Metal tape embossing press; British Automatic Co. Ltd.
A861 Modern dial-type coin-operated personal weighing machine; British Automatic Co. Ltd., 14 Appold Street, London E.C.2.
A862 Henry Booth and Universal Time.
A866 Original copy of the first 'Bradshaw'; The Executors of the late W. H. V. Bythway, Solicitor, Pontypool, Mon.
A867 Portrait of George Bradshaw (1801-1853); Henry Blacklock & Co. Ltd., 5 Surrey Street, London W.C.2.
A868 Coghlan's 'Iron Road' book ' or 'The London, Birmingham & Liverpool Railway Companion' (c. 1838); Harold Wyatt, 8 Upper St. Martin's Lane, London W.C.2.
A869 Facsirnile copy of the first 'Bradshaw's Railway Guide', published 19 Oct.1839 ; Bradshaw's Railway Guides, 5 Surrey Street, London W.C.2.
A870 Original copy of 'Bradshaw's Railway Timetable, 1845'; Bradshaw's Railway Guides.
A871 Copy of 'Bradshaw's British Railways Guide, 1951' ; Bradshaw's Railway Guides.
A872 Original Liverpool & Manchester Railway sheet timetable (1838); Harold Wyatt, 8 Upper St. Martin's Lane, London W.C.2.
A874 British Railways timetable (1951); The Railway Executive (British Railways).
A875 Midland Railway excursion poster (1851); Bemrose & Sons Ltd., Midland Place, Derby.
A880 Disc and crossbar signal, first brought into use on the G.W.R. (1840); The Railway Signal Co. Ltd., Fazakerley, Liverpool 9.
A881 Early form of semaphore, manually operated; The Railway Signal Co. Ltd.
A882 Lower quadrant semaphore signal with 'Adlake' signal lamp, manually operated; Lamp Manufacturing & Railway Supplies Ltd., 12-13 South Place, London E.C.2.
A883 Somersault semaphore signal, manually operated, with a 'Dutton' lamp; The Railway Signal Co. Ltd., Fazakerley, Liverpool 9; lamp by Lamp Manufacturing & Railway Supplies Ltd.
A884 Semaphore signal, motor-operated, 110 volts A.C., 2-position lower quadrant with 'Adlake' electric lamp; Westinghouse Brake & Signal Co. Ltd., Chippenham, Wilts.; lamp by Lamp Manufacturing & Railway Supplies Ltd., 12-13 South Place, London E.C.2.
A885 Semaphore signal, motor-operated, model 2A, 10 volts D.C., 3-position upper quadrant with electric lamp; Metropolitan-Vickers-G.R.S. Ltd., 132-135 Long Acre, London W.C.2.
A886 Miniature model of Stevens & Son's 'Hook-Lock' frame (c. 1859): one of the early attempts at modern interlocking; Westinghouse Brake & Signal Co. Ltd., Chippenham, Wilts.
A889 Model of McKenzie & Holland's 'Soldier-Locking' frame, patented 1873 ; The Science Museum, South Kensington.
A890 Model of Cam and Tappet frame, patented 1888 ; The Science Museum, South Kensington.
A891 4-lever ground frame, interlocking provided if required (not shown); Westinghouse Brake & Signal Co. Ltd., Chippenham, Wilts.
A892 Original Cooke & Wheatstone double needle electric telegraph (1837); The Railway Executive (British Railways).
A896 Preece's block-signalling instrument (c.1862), introducing miniature semaphore arm to indicate 'Line Clear' or 'Line Blocked' ; The Science Museum, South Kensington.
A897 Siemens' block bell, magnetic induction type; The Science Museum, South Kensington.
A898 Tablet instrument of the first type, for signalling and protection of railway movements on single lines (c 1870); Tyer & co. Ltd., Ashwin

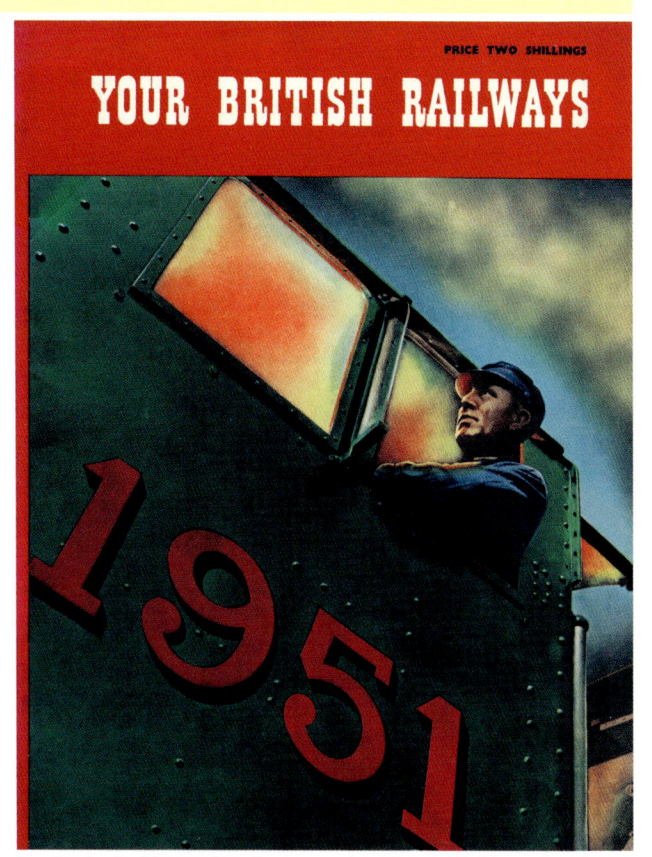

For the Festival BR produced a 48 page booklet entitled 'Your British Railways' highlighting the developments since 1948 but very much still steam orientated. *Images courtesy of Mike Ashworth*

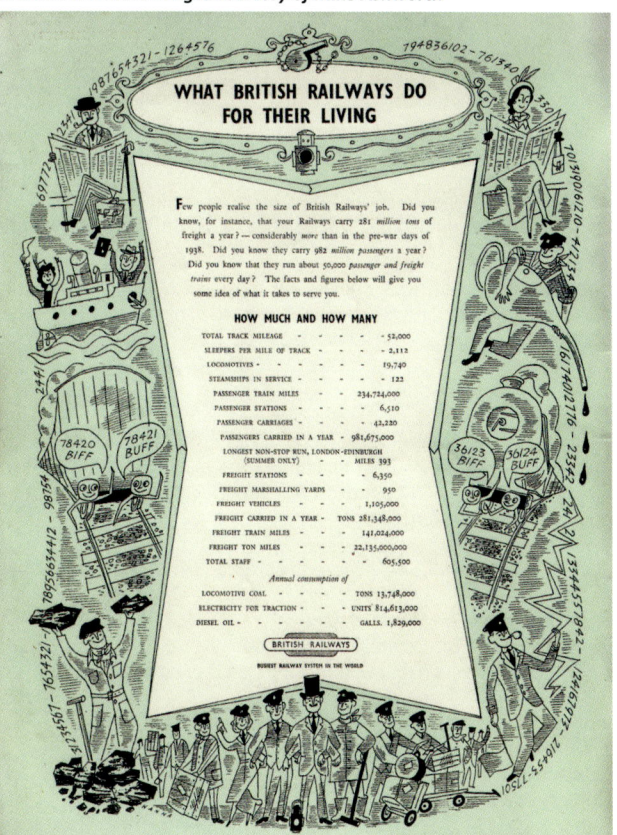

RAILWAY TIMES 1951

Street, Dalston, London E.8.
A899 Webb-Thompson large type staff instrument for the control of traffic on single-line railways, first brought into use 1889; The Railway Signal Co. Ltd., Fazakerley, Liverpool 9.
A900 Electric interlocking block instrument for controlling traffic in conjunction with track circuiting; W. R. Sykes Interlocking Signal Co. Ltd., 26 Voltaire Road, Clapham, London S.W.4.
A901 Key token instrument, for signalling and protection of railway movements on single lines, 1946 pattern ; Tyer & Co. Ltd., Ashwin Street, Dalston, London E.8.
A902 Webb-Thompson miniature staff instrument, developed in 1904, reducing size and weight of staff; The Railway Signal Co. Ltd., Fazakerley, Liverpool.
A903 The electric token principle in railway signalling.
A906 Electro-pneumatic point operation employing compressed air 60/80-lb pressure per sq. in., with anti-freezing devices if required (not shown); Westinghouse Brake & Signal Co. Ltd., Chippenham, Wilts.
A907 All-electric point machine, model 5A, 110 volts D.C: complete operating mechanism and detector in one unit; Metropolitan-Vickers-G.R.S. Ltd., 132-135 Long Acre, London W.C.2.
A908 Type HA electric point machine;. The Siemens & General Electric Railway Signal Co. Ltd., East Lane, Wembley, Middx.
A909 3-aspect searchlight signal, with position light junction indicator, A.C. or D:C. operation, light-weight aluminimum case, fourth aspect added if required (not shown); Westinghouse Blake & Signal Co. Ltd., Chippenham, Wilts.
A9I0 Multi-lamp route indicator, type ML long range, single panel, indications as required, framed by lamps in series, 110 volts A.C. ; Metropolitan-Vickers-G.R.S. Ltd., 132-135 Long Acre, London W.C.2.
A911 Type LA long range colour light 4-aspect signal ; The Siemens & General Electric Railway Signal Co. Ltd., East Lane, Wembley, Middx.
A9I5 Short range colour light shunt signal type, ME, 2-aspect, 6⅜-in doublet spread-light lenses, 110 V. lamps ; Metropolitan-Vickers-G.R.S. Ltd., 132-135 Long Acre, London W.C.2.
A916 Solenoid-operated disc shunt signal, with flood lighting unit, A.C., for ground or post mounting Westinghouse Brake & Signal Co. Ltd., Chippenham, Wilts.
A917 Type LF position light shunt signal; The Siemens & General Electric Railway Signal Co. Ltd., East Lane, Wembley, Middx.
A9I8 'Westlyte' combined control panel and illuminated diagram, with plastic 'piped' lighting providing the temporary and permanent indications; Westinghouse Brake & Signal Co. Ltd., Chippenham, Wilts.
A920 Fog signalling apparatus (c.1898) actual specimens of apparatus used on the line; The Science Museum, South Kensington, London S.W.7.

The York exhibition was opened by the Lord Mayor of York with an introduction by H A Short Esq. CBE MC Chief Regional Officer NE Region. Amongst the items featured were Locomotion No. 1, Stockton & Darlington Coach No. 3, Queen Adelaide's coach and Queen Victoria's saloon.

Back in London a number of locomotives in addition to the Indian Railways exhibit previously mentioned were exhibited at the Festival including the following from BR which were located in a display area adjacent to Hungerford Bridge –

No.70004 *William Shakespeare* finished in special exhibition condition
No 10201, 1Co-Co1 diesel electric
No 26020 , EM1 Bo-Bo electric later Class 76
A Mk 1 coach section showing half a third class compartment

Industrial locomotives were also represented by:
Shutt End Colliery (Foster & Raistrick) 0-4-0 *Agenoria*
Guiness Brewery (Hibbert) 4wDM No. 36
ICI Osborne (Ruston & Hornsby) 0-4-0 DM *W L Raw*

BR's North Eastern Region also staged an 'Exhibition of Railway Rolling Stock' at York old station as part of York's Festival Fortnight. It ran from June 4 - 16 1951 with the programme priced at 3d.

London Transport exhibited :
New 'R49' Underground stock in aluminium with an unpainted section to reveal construction

A foreign exhibit brought over from France was also included in the shape of –

Buddicom 2-2-2 No. 33 (running as classmate No. 3)

Locomotives illustrative of the export drive were also present –

English Electric Bo-Bo for the Tasmanian Government Railway 3ft. gauge
Ruston & Hornsby 0-4-0 DM165 hp 5' 3" gauge

Finally one should perhaps mention the 'other railway' to be found at the festival site in Battersea Park in the shape of Rowland Emett's creation the 'Far Tottering and Oyster Creek Railway'. Emmett who was a cartoonist for Punch magazine, was noted for his illustrations of bizarre forms of transport. A 15" gauge line ¾ mile in length was opened in the park to coincide with the Festival and featured three steam outline diesel locomotives named *Wild Goose*, *Nellie* and *Neptune*. Hugely successful in spite of a fatal accident shortly after it opened, caused by the three locomotives operating simultaneously, it outlived the Festival moving to another section of the park until 1975.

A couple of fascinating views of No 10202 being jacked over into position for the Festival under the watchful eyes of Motive Power Inspector Richford and Fitter Mitchell. In the background can be seen the Shot Tower of the former Lambeth Lead Works which was a prominent landmark featuring in a number of paintings including those of J M W Turner. In 1950 the gallery chamber at the top of the tower was removed and a steel-framed superstructure was added as seen here thus providing a radio beacon for the Festival. After closure of the Festival the tower was the only building to be retained on the site but in 1962 it was demolished to make way for the Queen Elizabeth Hall which opened in 1967. In the background to the second image is the 'Skylon' the futuristic metal sculpture symbolising the bright new future which the Festival embodied. The locomotive manoeuvres took place near the site of the former turntable located on the up side of the Hungerford Bridge approaches. This was once part of the former Belvedere Road depot with the turntable being used at one time by locomotives terminating at Charing Cross. *Lens of Sutton, Dennis Culum collection LSDC1166/7*

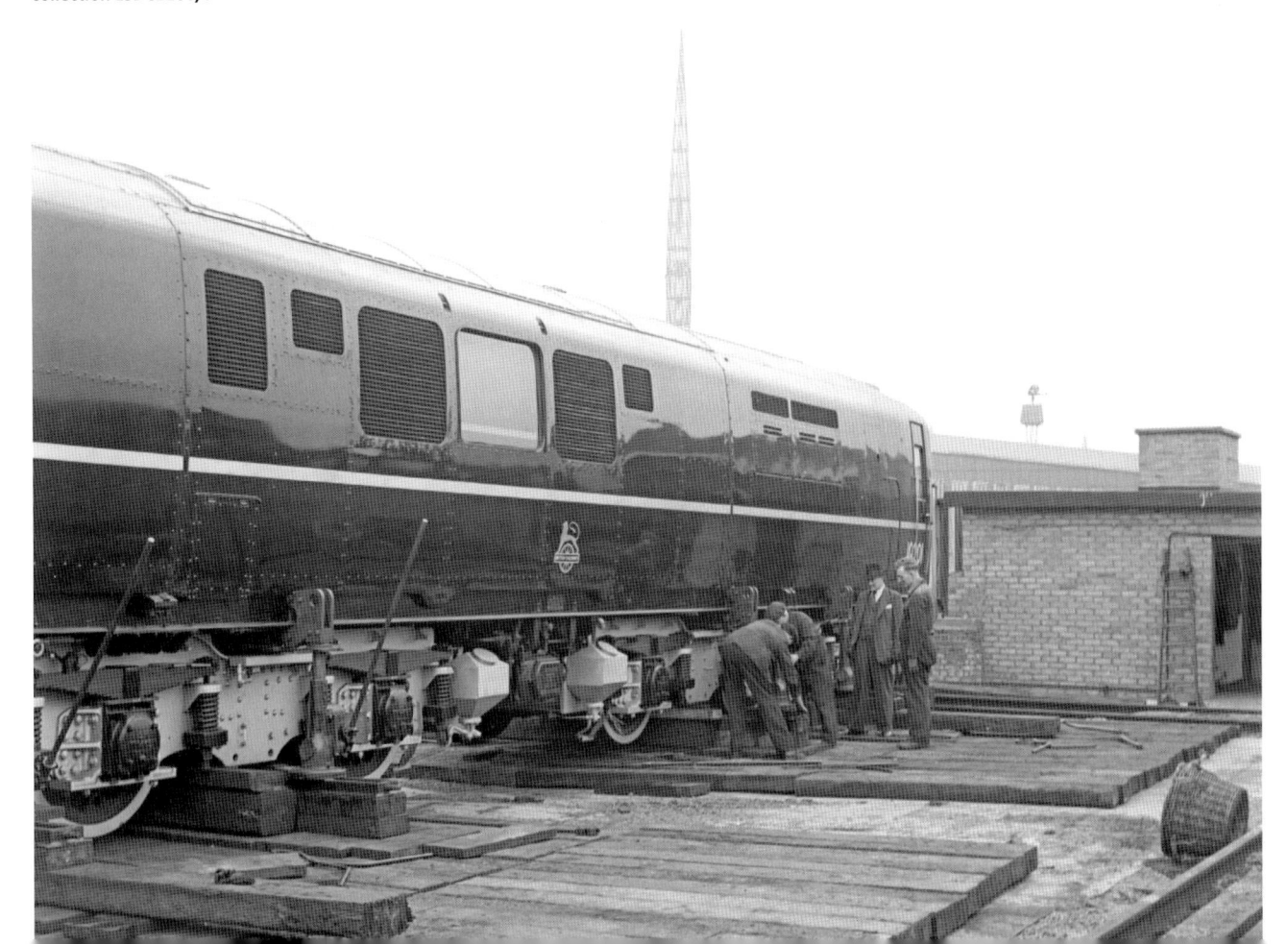

Brief Encounters

OLD SOLDIERS NEVER DIE. Above: Standing in the open coupled to large pipes and with very tall chimney extensions two GNR Ivatt Atlantics, Nos 3274 and 3285, continue to provide steam for Doncaster Works. They were withdrawn from service in May 1946 with both being finally cut up in March 1952. The frame and shell of classmate No. 4461, withdrawn in August 1945, was still intact at Doncaster five years later in November 1950.
Neville Stead NS202342 / Transport Treasury

WELL SIX AND A HALF MILES NEARER! Opposite: South Molton Road station on the Exeter – Barnstaple line was finally renamed King's Nympton on 1 March 1951. This had the effect of reducing the mileage from the station to its namesake settlement from 9 miles, to South Molton, to just 2½ miles, to King's Nympton. South Molton had a much more convenient station situated on the Taunton – Barnstaple line which had been in business since 1873. But of course this was a GWR station which the LSWR would not be keen to acknowledge.

TEN BOB A JOB

So acute is the manpower shortage on BR in certain areas that an incentive is being offered to existing staff who introduce a new recruit. The scheme is restricted to those occupations where the shortage is most keenly felt i.e. motive power, traffic, civil engineering and S&T departments and will run provisionally until October 1951 after which it will be reviewed in the light of take up. 10/- (50p) is on offer to existing staff when they secure a new member of staff with a further 10/- on offer should the recruit manage to complete two month's service without deciding that a life on BR is not for him. There has been excessive wastage in the past when new staff leave after a short time and it is hoped this scheme will go some way towards alleviating this. Reliance has been placed on existing staff being sufficiently motivated to paint an attractive picture of employment with BR. (Ed: 10/- in 1951 is the equivalent of £20 in 2024). The manpower shortage resulted in a drastic reduction in services for the winter timetable the introduction of which was brought forward by two weeks.

RAILWAY TIMES 1951

Viewed from the down platform looking north west, the signalbox and station building at Kings Nympton are captured basking in the summer sun in this timeless scene of a sleepy wayside station – now of course much nearer to its advertised location. Lens of Sutton, *Dennos Culum colelction LSDC 2636*

The town of South Molton had its own station, much nearer than the SR one, which is here visited by 43xx class No 7306 en route to Barnstaple in 1964. *RP 1198*

WOODHEAD BORES MEET. On 16 May 1951 it was reported that work had been completed on the 12 foot square pilot headings from both ends of the new tunnel through the Pennines linking Woodhead and Dunford Bridge and from the foot of the ventilation shaft some 467 feet deep located near the middle of the bore. The two headings joined some 1 mile from the Woodhead entrance. Work then proceeded to enlarge the bores to the required size of 24 feet high and 31 feet wide lined throughout with concrete to accommodate the double track. Apart from a 600 foot 40 chain curve at the Woodhead end the tunnel which is over 3 miles long will be on a straight alignment. Field telephones maintain contact between the workface and the works offices. Work was planned to finish in 1952 by which time more than 1m tons of rock will have been excavated from the approach cuttings and the tunnel itself. Work on this new bore was put in hand after it became clear that the twin single line bores, dating from 1839-52, had become uneconomic to maintain. At Dunford Bridge the existing down side station buildings will be demolished to make way for the approach line to the new tunnel.

RAILWAY TIMES 1951

THE 'BRITANNIA' INCIDENT. Jubilee class No 45700 *Britannia* has recently undergone a name change to *Amethyst* thus permitting a new Standard pacific No 70000 to carry the name *Britannia*. No 45700 was named upon completion in April 1936 but now bears the name of the frigate H.M.S. Amethyst which achieved notoriety running the gauntlet of heavy fire from the People's Liberation Army during a passage down the Yangtze river on 29 April 1949 this becoming known as the 'Yangtze Incident'. The vessel was trapped in China until 30 July 1949 when she escaped under cover of darkness. In 1957 a film was made of its escapade starring Richard Todd and the ship was brought out of reserve to play herself. As the engines were no longer operational, sister ship H.M.S. Magpie was used for shots of the ship moving. The Suffolk rivers Orwell and Stour stood in for the Yangtze. Whilst the frigate went to the breaker's yard shortly after filming was completed, the 'Jubilee' soldiered on until withdrawal in July 1964 and of course *Britannia* is still with us.

Top: No 45700 *Amethyst* double heading with an unidentified locomotive speeds past Heaton Lodge in this undated view. *Neville Stead NS203786 / Transport Treasury*

Left: No 45700 *Amethyst* receives attention inside Polmont shed. *R E Vincent REV96C-2-5 / Transport Treasury*

RAILWAY TIMES 1951

NO SHOW AT THE END OF THE PIER. Two stations at piers have recently closed to passenger traffic. Port Victoria on the Hundred of Hoo branch shut its doors on 11 June 1951 when the land was required by the adjacent BP oil refinery and passenger services on the branch were truncated at a new station at Grain. The 400 foot pier which stretched into the River Medway estuary had been shortened to just 93 feet back in 1916 when the seaward end was declared unsafe and barricaded off. By 1931 the whole pier was declared unsafe and a new station built at the landward end. In 1941 the remnants of the pier were demolished and just two trains daily were provided which was apparently newsworthy enough to justify the filming of two Pathe News items, in 1939 and 1947, both featuring the less than onerous duties of the Station Master for as the commentary of one film states, "Four tickets to collect for one train is a rush, eight visitors a day is a crowd".

Right: An R class 0-4-4T No 31659 waits at Port Victoria on 3 October 1950 with a returning service to civilisation at Gillingham. This veteran built in 1891 by Sharp Stewart would be withdrawn the following year. *Neville Stead NS200464 / Transport Treasury*

The other pier station closed this year, from 2 July 1951, was Felixstowe Pier at the end of the branch connecting the town with the GER mainline at Westerfield. Situated in the dock area of the town it was not the most attractive location but this did not prevent BR from locating a couple of six berth Camping Coaches there after closure from 1952 until 1954 when the number increased to four coaches until 1959 when they moved inland to the Town station in Felixstowe.

THE BARD'S SEIZURE. The failure of Britannia class No 70004 *William Shakespeare* whilst working the up Golden Arrow on Sunday 21 October just ten days after its initial appearance on this prestige service has led to the temporary withdrawal of the whole class of pacifics pending investigations into the causes of the coupling rod breakage which led to this failure. No. 70004 ground to an unscheduled halt at Headcorn with its motion seized rendering the locomotive immovable until fitters arrived which led to considerable delay for boat train passengers. The disgraced locomotive was hauled to Ashford Works where it sat awaiting the delivery of new parts from its birthplace at Crewe Works. It is understood that the problem related to the wheels shifting on their axles. The driving wheels were modified and new plain front coupling rods fitted after which the locomotive returned to service on 7 December when it hauled the 9.00 am boat train from Victoria to Dover. Apparently few services rostered for Britannia haulage on other regions of BR were affected by this temporary embargo on the class although, as mentioned in another article in this volume, the Eastern region did receive some Bulleid pacifics whilst the Britannias were temporarily unavailable. The 'Brits' gradually returned to traffic with eight expected to be available before Christmas 1951 and the remainder shortly afterwards. Here we see happier times with No 70004 *William Shakespeare* running into Folkestone Junction with the down 'Golden Arrow'. *Neville Stead NS207731 / Transport Treasury*

In the next issue covering British Railways in 1952

Turbomotive No 46202 is seen at Crewe. It was rebuilt in 1952 as a non-turbine driven locomotive and named *Princess Anne* but the pacific operated for only two months before it was wrecked in the Harrow & Wealdstone disaster of 8 October 8 1952 and subsequently scrapped. *Neville Stead NS203427, Transport Treasury*

The frames of No 46202 following the Harrow disaster. Note that the namplate is still in position. *R E Vincent REV73C 6-4, Transport Treasury*

Amongst the items it is intended to feature the following –

HARROW & WEALDSTONE DISASTER

RISING OF THE CLANS

SHORT LIVED REBUILDING OF THE TURBOMOTIVE

WOODHEAD ELECTRIFICATION STAGE 1

VECTIS CLOSURES

WHITSTABLE'S WINKLE LINE CLOSED

SWINDON'S 2-6-2 STANDARD TANKS

ACV RAILBUS

FRANCO & CROSTI